ATOMIC FORCE MICROSCOPY

A Concise Introduction

ATOMIC FORCE MICROSCOPY

A Concise Introduction

Nancy A Burnham

Worcester Polytechnic Institute, USA & ETH Zurich, Switzerland

World Scientific

NEW JERSEY • LONDON • SINGAPORE • BEIJING • SHANGHAI • HONG KONG • TAIPEI • CHENNAI • TOKYO

Published by

World Scientific Publishing Co. Pte. Ltd.
5 Toh Tuck Link, Singapore 596224
USA office: 27 Warren Street, Suite 401-402, Hackensack, NJ 07601
UK office: 57 Shelton Street, Covent Garden, London WC2H 9HE

British Library Cataloguing-in-Publication Data
A catalogue record for this book is available from the British Library.

ATOMIC FORCE MICROSCOPY
A Concise Introduction

ISBN 978-981-98-2309-3 (hardcover)
ISBN 978-981-98-2430-4 (paperback)
ISBN 978-981-98-2310-9 (ebook for institutions)
ISBN 978-981-98-2311-6 (ebook for individuals)

For any available supplementary material, please visit
https://www.worldscientific.com/worldscibooks/10.1142/14578#t=suppl

Desk Editor: Rhaimie B Wahap

Typeset by Stallion Press
Email: enquiries@stallionpress.com

Dwight C Burnham (1922–2016), for stimulating my curiosity about the universe

Foreword

I have had the pleasure of knowing Nancy Burnham since the early 'nineties, when she was working as a senior researcher at the EPFL in Lausanne and I had just started as a professor at the ETH Zurich. As academic neighbors with a strong interest in AFM, we rapidly started a project together that lasted until her return to the USA.

Since I have known her, Nancy has been my go-to person on AFM fundamentals. She was in the field from the very beginning, having worked with Rich Colton at the Naval Research Laboratory in Washington DC, where many of the important early AFM developments had taken place. Over the years I have consulted her many times, often on niche aspects of AFM measurements and always received either the definitive answer or an invaluable pointer.

Soon after beginning her professorship at Worcester Polytechnic Institute (WPI) at the start of the new millennium, Nancy began the development of a comprehensive AFM course that teaches the subject, all the way from fundamentals to the practical use of an AFM instrument. One of the spin-offs from the course was an extensive set of YouTube videos that cover both the fundamentals and actual instrument operation. In 2015 I was delighted to be able to convince Nancy to bring her course to my department at the ETH, in the form of a two-week summer intensive workshop for graduate students. We jointly taught the course until my retirement (Nancy as professor and me as a glorified teaching assistant!), my successor Lucio Isa having taken over my role since my retirement in 2020.

The form and content of Nancy's course were a revelation to me, who had only taught AFM in a classical (probably boring) 2h lecture prior to that time. The first aspect that impressed me was the efficient use of the "inverted classroom": Over an intensive two weeks she expected that the students watch a YouTube video from her extensive AFM series prior to each day of the course. Thus, rather than starting with a dry lecture, each day began with a quiz on the video and the opportunity for the students to ask questions on the video content, to ensure that everybody was clear about the physics behind the practical part of the day that was to follow.

We were lucky enough to have acquired three AFMs from a Swiss manufacturer that were similar to those used in Nancy's course at WPI. We therefore could give the students many hours of practical time with the instruments during the course.

The students worked in groups of three, which seemed to give each one plenty of opportunity to gain hands-on experience. The class was limited to a maximum of 18 students, so at any one time, half of them were going through carefully planned mesurements with the AFMs while the other half were alternating between software exercises to analyze AFM data and macroscopic experiments (which I had the pleasure of supervising). These were intended to convey the mechanical aspects of the AFM, but on a more familiar size scale! They involved dealing with steel cantilevers, weights and magnets (the macro-world's answer to van der Waals' forces), and the measurement of spring constants via macroscopic displacements and vibrations.

Over the last few days of the course, having acquired the necessary skills to use the instruments in several different modes with some confidence, the students carried out a mini, AFM-based research project, often related to their PhD research. This was reported in a final project presentation on the last day of the course. The 18 places that we made available were generally filled within a week of opening enrollment each year, and now nearly 200 graduate students have received this excellent grounding in AFM which, for many, has been central to their doctoral work.

I am delighted that with this textbook, Nancy's innovative and effective approach to teaching AFM will now be available to everybody. Readers can become confident in their knowledge of both the fundamentals and the practical use of an instrument that has become central to a vast range of research fields in the physical, chemical, and biological sciences.

Prof. em. Dr. Nicholas D. Spencer
Department of Materials
ETH Zurich
Switzerland

August 2025

Preface

This textbook originated in courses on Atomic Force Microscopy (AFM) at Worcester Polytechnic Institute (WPI); the first rendition of the course was in 2001. Targeted towards second-year undergraduates in science and engineering, it attracted learners from first-year students to retired professionals. It was clear that the skill levels of such a wide range of people varied widely, motivating the course's subsequent split into undergraduate and graduate versions. The undergraduate version, PH 2510, is taught over seven weeks, corresponding to one of WPI's undergraduate terms. The graduate version, PH 561, is taught over a fourteen-week semester, which graduate students find easier to accommodate their other responsibilities. Starting in 2015, I led the teaching of the material to predominantly graduate students in Materials Science at ETH Zurich. This block course runs outside of the semester in a two-week, all-day, every-day format.

While the fundamental AFM-related content is the same for all three courses, I vary the level of writing and projects expected. For the longer courses at WPI, I require six formal instrument lab reports, with the aim to prepare students for scientific communication at the level of a draft of a manuscript for submission to Applied Physics Letters. Because the undergraduate version of the course is time-restricted compared to the graduate version, there is no room in the syllabus for a laboratory-based project. Instead, undergraduates are asked to give short talks on two related papers from the AFM literature. The graduate students, on the other hand, have a few weeks to conduct a lab project on a topic of interest to them. For the two-week block course at ETH Zurich, the emphasis on writing is put aside, yet two three-hour lab sessions are scheduled for a "nano" project, with the morning of the last day devoted to student presentations.

I chose to cover fewer topics at a fundamental level, rather than many topics at a graduate or higher level: there are many, many AFM modes of operation and a vast, vast literature that I omit so as to concentrate on the basics of AFM within about two hundred pages—concisely yet comprehensively. The intent is to educate the curious to competently use AFM and provide them a firm foundation for further study, be they students in a course or independent learners.

Although perhaps a dozen books on AFM have been previously published, to my knowledge, this is the first *textbook* on AFM, with learning objectives, suggestions for further reading, and problems at the end of each chapter. The reading lists are short, so as to not overwhelm the reader. The suggested readings fall into one of three categories: influential or illustrative historical publications, tutorial-style works, or "full-story" books. I also include a few of my own relevant contributions; the full list is at Google Scholar and Orcid.org. (The latter is less frequently updated.)

The appendices provide compiled learning objectives (Appendix A), advice for conducting experiments (Appendix B), a description of the electronic supplement available free to interested readers (Appendix C), and lists of acronyms and symbols (Appendices D and E). These are followed by the answers to the end-of-chapter problems, which should be helpful to independent learners. The final section is an index. Within the text, I note the relevance of complementary material from my YouTube/AtomicForceMicro channel or from the supplement by means of boxed or highlighted stand-alone sentences.

The background assumed for this textbook is first-year physics and calculus, except for the last chapter on oscillators, for which an understanding of differential equations and complex numbers is needed. Some introductory physics and materials science is incorporated within Chap. 5. The perspective is that of a physicist—readers might notice some subtle and not-so-subtle physics lessons! The goal is to ensure that both students and independent learners are well prepared for their research, be it senior capstones, graduate studies, or professional work.

At WPI, students with majors in the sciences (physics, applied physics, mathematics, computer science, environmental science) and engineering (biomedical, chemical, electrical), have all prospered in the course, as well as those with minors in physics, materials engineering, and nanoscience. At ETH Zurich, the predominant group of enrollees has been Masters students in materials science, as well as a smattering of PhD students in materials, chemistry, and civil engineering.

This textbook is organized into two parts. The first part covers fundamentals of AFM: an introduction to AFM principles, the challenges of imaging, an overview of related instruments, and the more common AFM modes of operation, and AFM design and calibration. The second part concentrates on a mode of data acquisition that requires interpretation based on materials physics, that of force-curve acquisition and its subsequent analysis. Along the way, some basics of physics and materials science are explained, as well as different kinds of intermolecular and contact forces that can act between the AFM tip and sample. Finally, in the last chapter intended for intermediate-level or higher science and engineering students, models for cantilever oscillation are presented.

I hope this textbook will elevate readers' AFM skills and open the door to fascinating future studies of the nanoscale.

Professor Nancy A Burnham, PhD
Department of Physics, Worcester Polytechnic Institute, USA
Department of Materials Science, ETH Zurich, Switzerland

June 2025

Acknowledgements

My AFM courses would not have been possible without the prompt of my department head, Professor Thomas Keil, after I first arrived at Worcester Polytechnic Institute (WPI) in 2000. I am indebted to him and to the Physics Department for their continuing support.

I acknowledge a 2004 grant (# 0406687) from the US National Science Foundation for a term off from teaching at WPI near the beginning of the writing process.

Since 2013, I have been using Nanosurf's NaioAFMs for teaching. I am indebted to the Nanosurf teams in Woburn, Massachusetts and Liestal, Switzerland for help in keeping them in good repair.

I am grateful to the Department of Materials at ETH Zurich for being able to teach AFM in a block-course format starting in 2015. In particular, Professor Nicholas Spencer, Professor Lucio Isa, and Dr. Shivaprakash Narve Ramakrishna helped innovate and teach the laboratory sessions with a macroscopically sized cantilever. Dr. Lorenzo De Pietro provided his expertise with ETH's learning management system. And Ms. Josephine Baer expertly coordinated my stays.

Moreover, throughout the years there have been many students and teaching assistants at both WPI and ETH who—through their questions and insights—have made this a better tome.

I am deeply pleased that Mr. Christopher B. Davis of World Scientific Publishing approached me in 2022. He provided encouragement and valuable suggestions during the last sustained push to complete the body of this work.

Finally, I very much appreciate the sharp eyes of my colleagues Prof. Diana Alatalo, Dr. Nico Kummer, and my husband Mr. Frederick Hutson for their insightful edits and suggestions as the manuscript neared completion.

Contents

Part I
Instrumentation

Chapter 1
Context and overview of instrumentation

Abstract This first chapter introduces atomic force microscopy (AFM) as a fascinating gateway nanoscale characterization technique. It explains the operational principle of AFM and describes its advantages and disadvantages. The purposes and plan of the book and accompanying courses are listed, and the typical hardware and software of AFM are described. As for all of the chapters, it concludes with learning objectives and problems.

1.1 Introduction

Why is atomic force microscopy so compelling? At least two reasons.

First, using a simple principle, we large-scale humans have access to an incredibly tiny scale, on the order of nanometers, the size scale of atoms and molecules. To grasp how small a nanometer is, imagine that we can blow it up to the size of a marble, about 1 cm in diameter. Then, to remain in proportion, a meter would be 10,000 kilometers, a little less than the distance between New York City and Tokyo. That's a lot of marbles!

Second, atomic force microscopy (hereon, AFM) is a gateway tool. Its use opens many doors to the captivating science and technology of the very small, known as nanoscience and technology, or alternatively nanotechnology. Whereas the computer industry has used a "top-down" approach to make ever smaller features on computer chips, nanotechnology frequently aspires to understand "bottom-up" processes, such as those that cause cells to organize into living beings or how cancer cells might be induced to ingest cancer-killing nanoparticles.

At the nanometer scale, science and technology become very interdisciplinary. Through AFM, a researcher is exposed to and can learn from adjacent fields of physics, chemistry, biology, and engineering. Conversely, researchers can find themselves making significant contributions to fields outside of their original training. AFM is a means for scientific give-and-take.

1.1.1 AFM operational principle

An atomic force microscope works much like you do when you are searching for a light switch on a wall in a darkened room. Your fingers scan back and forth on the wall until you feel the switch. Similarly, in an AFM, a flexible “arm" (cantilever) has a sharp “finger" (tip) on its end that detects features on a surface by deflecting as the cantilever and sample are scanned relative to each other. The deflection is observed by a laser or similar device that bounces off the back side of the cantilever. The reflected laser light hits a position-sensitive photodiode that turns the position of the laser spot into a voltage that is then recorded by a computer; see Fig. 1.1a. In this book, the acronym “AFM" is used interchangeably for “atomic force microscopy" and “atomic force microscope."

Figures 1.1b&c respectively show how the reflected laser spot moves up and down, revealing the cantilever's vertical motion, or back and forth, revealing twisting of the cantilever due to lateral forces from the sample surface. Both motions are possible simultaneously, such that the reflected light spot could move diagonally into one of the photodiode's four quadrants; each quadrant outputs a voltage that is proportional to the amount of light upon it.

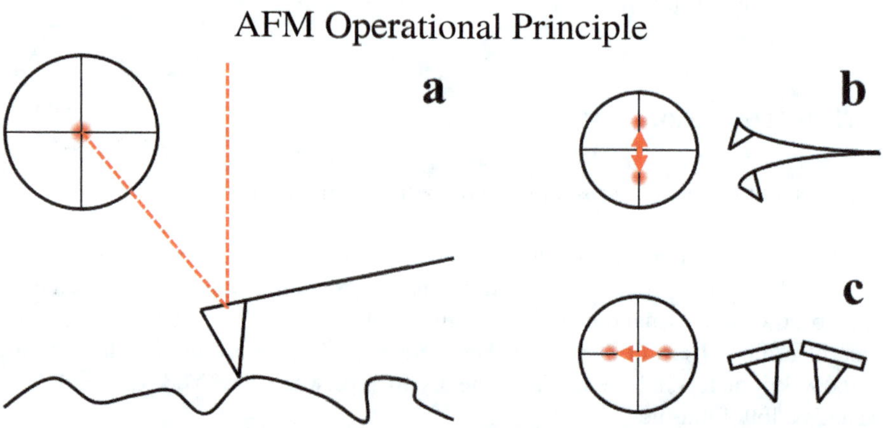

Fig. 1.1 A simplified sketch of AFM's operational principle. (a) Motion of the flexible cantilever with a sharp tip near the sample surface is detected by a laser (dashed line) and quadrant photodiode. (b) Up-and-down motion of the cantilever results in up-and-down motion of the reflected light. (c) Twisting of the cantilever from lateral forces (e.g. friction) from the sample results in back-and-forth motion of the reflected light.

Of course, AFM is more complicated than this simple idea, and we will explore the details of a typical AFM setup in Sect. 1.2. Right now, let us discuss *why* (or why not) you might want to use an AFM.

AFM Lessons 1–2 at youtube.com\AtomicForceMicro

AFM Lessons 1–2 provide an introduction to AFM, with examples.

1.1.2 Disadvantages and advantages of AFM

Every instrument has its disadvantages and advantages compared to other techniques. Most research labs will have tools that complement AFM, the most common of which are optical and scanning electron microscopes. Here is a quick comparison among them.

Optical microscopes are standard tools in many laboratories. Perhaps the reader has used one to examine, for example, bacteria in a drop of pond water or a dead insect. Images in optical microscopy are formed essentially instantly because of the high speed of light. Furthermore, light comes into the eyepiece or camera from all viewable parts of the sample surface simultaneously. In comparison, the AFM tip or electron beam of a scanning electron microscope must raster over the sample in order to form an image. However, the resolution of optical microscopes are limited by the wavelength of visible light, which is in the range of 300-700 nm. Only the very best optical microscopes would be able to distinguish features in that size scale; more common optical microscopes are able to distinguish features only in the several-micron range.

Scanning electron microscopes (SEMs) operate by electrons being extracted from a source with a high electric field, focusing the now vertical electron beam using the equivalent of optical lenses, and rastering that beam horizontally over the surface of a flat sample. The high-energy electrons impinge upon the sample—it must be a conducting sample—and some, now with lower energy, escape the surface. These are collected by a detector, and it is the detector signal that is plotted as a function of the horizontal position of the electron beam that forms an SEM image. Because the electron beam would dissipate in air, the beam and sample must be in a partial or full vacuum, adding expense and limiting the size of samples that can be studied. However, sample surfaces can be thoroughly cleaned by various techniques within the vacuum, and the electron beam can be manipulated very easily and quickly by the “lenses," such that images are formed in real time. The researcher can quickly zoom from centimeter scan ranges to much, much smaller ranges that are easily sub-micron. The best SEMs have resolutions of a few nanometers, although more commonly the resolution is some tens of nanometers.

What are the *disadvantages* of AFM as compared to optical and electron microscopy? It is relatively slow, in that data collection requires mechanically moving the tip relative to the sample, which takes time compared to an electron beam, or especially, light. The magnification range is limited, typically starting at some tens or perhaps a hundred microns, down to submicron, instead of starting at the centimeter scale down to submicron. And compared to its horizontal motion on the order of tens of microns, the vertical range of most AFMs is on the order of ten microns.

What about the *advantages* of AFM? Most important is that three-dimensional information is collected. Researchers can determine the height of the features on the surface without having to break the sample and view it from the side, as one would in optical and electron microscopy. And similarly to optical microscopy, but different from SEM, there is minimal sample preparation and the medium can be liquids, air, or vacuum; SEM requires a conductive sample (of which many are not) and a vacuum. Furthermore, AFMs can do more than just image. The AFM tip can measure sample properties (as you would by pushing your finger into a material), move nano-objects across a sample surface (again, just like a finger), and AFM's cantilever technology can act as a measurement device for sensing (similar to your finger detecting the wind).

Please be aware that *each of these three classes of microscopes can damage your samples*. If the intensity of illumination in an optical microscope is too strong, the sample can be harmed. (Think of sunburn.) High beam energies and currents in SEM can change the chemical composition or even evaporate a sample. In AFM, damage can occur because of the very high pressure that can be created between the tip and sample; while forces are low, contact areas are miniscule, resulting in pressures that can easily change the shape of the tip or scratch the sample. Nonetheless, the capabilities of research labs would be severely hampered were it not for the ability of these three sorts of microscopes to determine micron- and nanometer-scale surface morphology.

1.1.3 Purposes and plan of this book and courses

More than one book could be written about AFM! The purpose of *this* book is to provide readers a firm foundation upon which they can learn more. It complements other materials for the undergraduate and graduate courses in AFM at Worcester Polytechnic Institute. The course has also been offered in a compact two-week block format at ETH Zürich. The initial version of the course was targeted to intermediate-level undergraduates in science and engineering who have completed their introductory courses in physics, which is the background assumed for this book's audience—no matter the age of the reader.

The complementary materials include YouTube videos (search for "Atomic-ForceMicro") and quizzes about their content, instrument lab instructions (introductory videos are at the YouTube site) and pre-lab quizzes, pedagogical software and instructions, exercises with a macro-sized cantilever, and in-class worksheets. Interested readers are welcome to request the complementary materials via the QR code at the end of Appendix C.

The objectives of the courses are:

- To become competent in the use of AFMs and in the interpretation of AFM data,
- To learn how physics applies to AFMs,

- To gain an understanding and appreciation of scientific research, and
- To develop the ability to write science well.

The overall goal of the course is not only to provide students with a valuable skill (AFM), but to help them develop their critical thinking and scientific communication skills such that they will be well prepared for a scientific or engineering career or to enhance their existing career.

Now that some context for AFM and this book and associated courses has been provided, in the next section, an overview of AFM instrumentation is given.

AFM Lesson 3 at youtube.com\AtomicForceMicro

AFM Lesson 3 provides an overview to the courses that form the basis for this textbook on AFM.

1.2 Instrumentation

An interesting aspect of AFM is how many different variations are possible. In this section, the focus is on the principles behind AFM hardware and software, so that the reader can quickly adapt from one AFM to another.

1.2.1 Hardware

The following subsections on hardware describe probes, the AFM as a system, and common variations on the standard AFM technique.

1.2.1.1 Probes

After learning about AFM's operational principle (Sect. 1.1.1), many people are curious about AFM cantilevers and tips. Together with the chip upon which they are fabricated, as well as any mounting sample, collectively they can be called the *probe assembly*. An AFM probe assembly is composed of (i) the *holder*, which eases the placement of the probe assembly into a specific microscope, (ii) the *chip*, to which the cantilever is attached, (iii) the *cantilever beam*, which bends in response to forces acting on it, and (iv) the *tip*, which makes contact with the sample. The chip might have multiple cantilevers on it. A typical probe assembly is sketched in Fig. 1.2, along with representative values of sizes and cantilever stiffness k_c.

Probes are made using the same techniques as for computer chips. Hence they are usually fabricated out of silicon or silicon nitride, two materials that are compatible

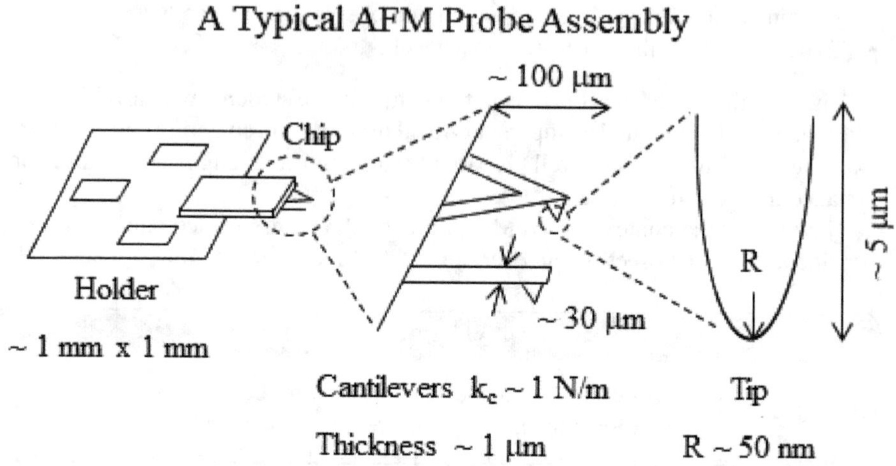

Fig. 1.2 Schematic of a typical AFM probe assembly. On the left is an overview of the holder, with the chip glued on top. (In other configurations, the chip might be inserted directly into an AFM.) In the center of the figure are shown two common forms of cantilever, a triangular one on top and a rectangular one on the bottom. The part of the probe assembly that actually touches the sample is the tip, seen on the right. Representative values for dimensions and spring constants are shown.

with computer-chip processes. A tremendous range of probe materials, sizes, shapes, tips, coatings, spring constants, and natural frequencies is available commercially to the consumer (you!). Why? Because each probe was designed with a particular application in mind. Indeed, a newcomer to AFM is often overwhelmed by the number of choices. In Chapter 4 an explanation of the thinking behind probe design is given, such that choosing probes might become easier.

The reader should be aware that everyday nomenclature for probes is loose; some people refer to the entire probe as a tip or a cantilever. Here, the word "tip" refers to that part of the probe that extends down from a cantilever to interact with the surface, and "cantilever" refers to the protrusion out from the chip that bends in response to forces upon it. The probe assembly, whether or not it is upon a holder specifically made for a particular AFM, is here referred to as a "probe."

● In **MacroLab 1**, you determine the spring constant of a macro-sized cantilever by two different techniques. ●

1.2.1.2 The AFM system

AFM Lesson 4 at `youtube.com\AtomicForceMicro`
AFM Lesson 4 is a video presentation of how an AFM works.

Naturally, there is more to an AFM than that shown in Fig. 1.1. In Fig. 1.3 the important components of an AFM are shown schematically. The now-familiar laser, probe, and photodiode are in the upper left. (Some AFM detection approaches use other techniques, but for brevity they are not described here. Nonetheless, they all measure the movement of the cantilever in some way.) A computer controls the scanner electronics that move the scanner. In this figure, the sample is shown on top of the scanner, but other approaches are to place the detection components (laser, probe, photodiode) on the scanner (typical for large samples), or to decouple the horizontal (x, y) and vertical (z) motions. In this last case, the sample is scanned in the x- and y-directions, with the detection on a vertical scanner.

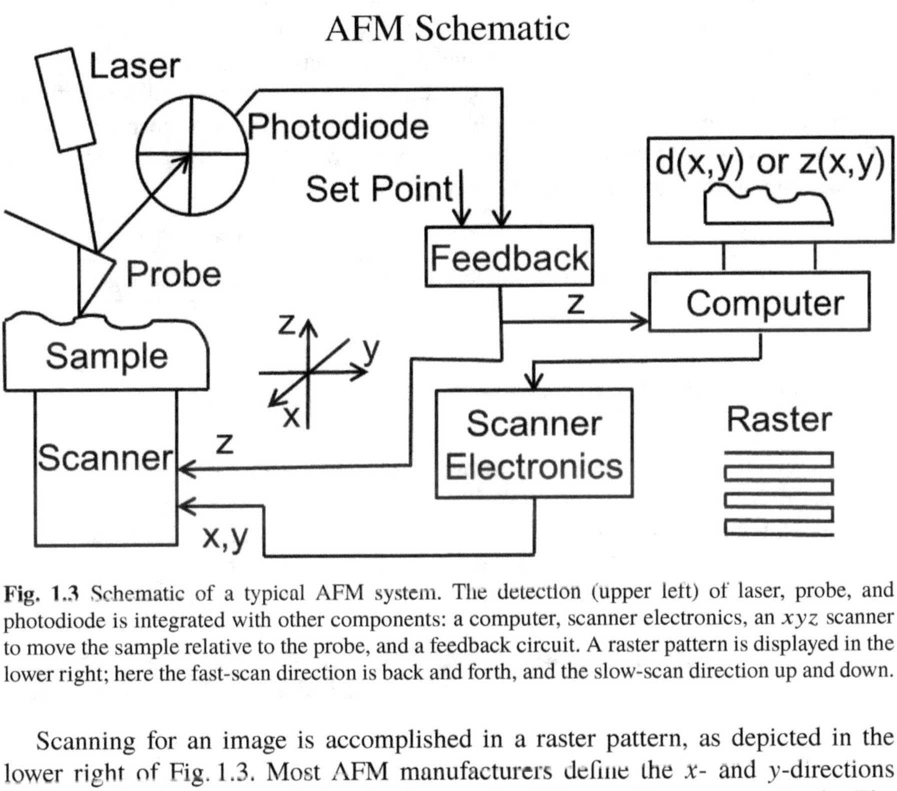

Fig. 1.3 Schematic of a typical AFM system. The detection (upper left) of laser, probe, and photodiode is integrated with other components: a computer, scanner electronics, an xyz scanner to move the sample relative to the probe, and a feedback circuit. A raster pattern is displayed in the lower right; here the fast-scan direction is back and forth, and the slow-scan direction up and down.

Scanning for an image is accomplished in a raster pattern, as depicted in the lower right of Fig. 1.3. Most AFM manufacturers define the x- and y-directions as perpendicular and parallel to the long axis of the cantilever, respectively. The default scanning option is to scan in x quickly and slowly move up or down in y. Because it can be useful to change the orientation of the raster (for example, to check for consistency of the image), the terms "fast-scan" and "slow-scan" directions are frequently used.

The photodiode voltage runs through a feedback circuit and is compared to a set-point voltage; the feedback circuit is important enough to merit its own subsection, Sect. 2.1. If the feedback circuit is not engaged, then the data that the computer collects are the deflection d of the cantilever as a function of horizontal position, $d(x, y)$, as the probe goes up and down over the sample surface and the z-position of the scanner does not change; this is constant-height mode. If the feedback is engaged (as shown in the figure), then the scanner includes z-motion while the scanner also operates simultaneously in x and y; this is constant-force mode. The amount of force is determined by the set point to the feedback circuit. Then an image is collected as $z(x, y)$ data.

Constant-height mode is faster than constant-force mode, since the feedback circuit takes some milliseconds to respond to changes in the photodiode voltage. However, constant-force mode holds the advantage that the cantilever's deflection, and in turn tip-sample force, is nominally constant while imaging, thereby limiting damage to the tip and sample. (From Hooke's Law, the magnitude of the force $F = k_c d$, with k_c the spring constant of the cantilever. Damage is caused by high pressure P due to the small area A of contact, $P = F/A$.) Moreover, interpretation of constant-force images can be easier when the force remains approximately constant.

The computer not only controls the scanner's x, y motion and collects deflection d or height z, it is the means by which image processing is performed, either in the same program as data collection or a separate one. Some of the common image processing options are described in Sect. 1.2.3.

In **Instrument Lab 1**, you learn how to acquire an AFM image.

1.2.2 The family of instruments

As presented earlier in Sect. 1.1.2, most research labs will complement AFM with optical and scanning electron microscopes. The reader should be aware that AFMs come in a wide variety of designs and functionalities; later in this book more details will be provided (Chap. 3). For the moment, in order to get you conversant in the field, a few terms and their acronyms will be introduced.

AFM Lesson 5 at `youtube.com\AtomicForceMicro`

AFM Lesson 5 covers scanning electron microscopy and compares it to AFM.

Historically, AFM has also been known as *scanning force microscopy* (SFM), but atomic force microscopy is now the more common description. AFM comes in many variations, depending on what the researcher wants to accomplish. If for example, the friction between the tip and a sample surface is of interest, the twisting of the cantilever is acquired during the raster scan; for a reminder see Fig. 1.1. This mode of operation is called either *friction force microscopy* (FFM) or *lateral force microscopy* (LFM).The theory behind this mode is outlined in Sect. 3.3. Many other AFM variations exist; to list them all would take up too much space in a book that attempts to lay foundations, not to be an up-to-date resource for the practitioner. Perhaps *you* will create the next innovation.

AFM is itself a variation of *scanning tunneling microscopy* (STM), in which the current between a sharp tip and a conducting sample is monitored. STM principles are explored in Sect. 3.2. Collectively, STMs, AFMs, and all their relatives are known as *scanning probe microscopes* (SPMs), or less commonly, *proximal probe microsopes* (PPM), or wryly, SXM, with the "X" standing for any variable. We slurp an "alphabet soup" of acronyms!

1.2.3 Software

This section provides a very brief look at AFM data acquisition software and the most common image processing options.

1.2.3.1 Data acquisition

The AFM's computer is loaded with AFM software. The software must interface with the user, who controls the parameters for data acquisition. It also interfaces with the hardware, and it must acquire, store, and display data. Most AFM software includes basic integrated image processing; some manufacturer's software is designed such that image processing is conducted through a separate program.

There is an enormous range in visual appearance of user interfaces. Frequently, the user interface is organized by panels, and an example of an image acquisition panel can be viewed in Fig. 1.4. It comprises four sections, two images on top (a,b) and two corresponding cross-sections on the bottom (c,d). The two panels on the left (a,c) display the data when the probe is moving left to right ("forward") over the sample, and the panels on the right (b,d) show the data when the probe is moving right to left ("backward"). The sample is an integrated circuit.

Figure 1.4 was made while the image was being acquired with the fast-scan direction in x and the slow scan in y proceeding from bottom to top. The scan range is 50 μm and the color scale (for which light colors are where the sample is higher) runs over nearly 3 μm. Note that the sample is mounted such that the images appear tilted in the x-direction; the left side of both images is darker than the right. Sample tilt is hard to avoid when one's instrument can measure microns and nanometers! Almost every image is leveled either during or after acquisition in order to clearly see the features of interest.

Regarding the cross-sections in Fig. 1.4c&d, one again notices the slope of the sample. The black lines relate to the active direction, and the gray ones the inactive one. So in (c) the forward cross-section is black, while in (d) the backward one is light gray. Please observe the shift between them. Causes for this hysteresis will be discussed in Sect. 2.1.2.

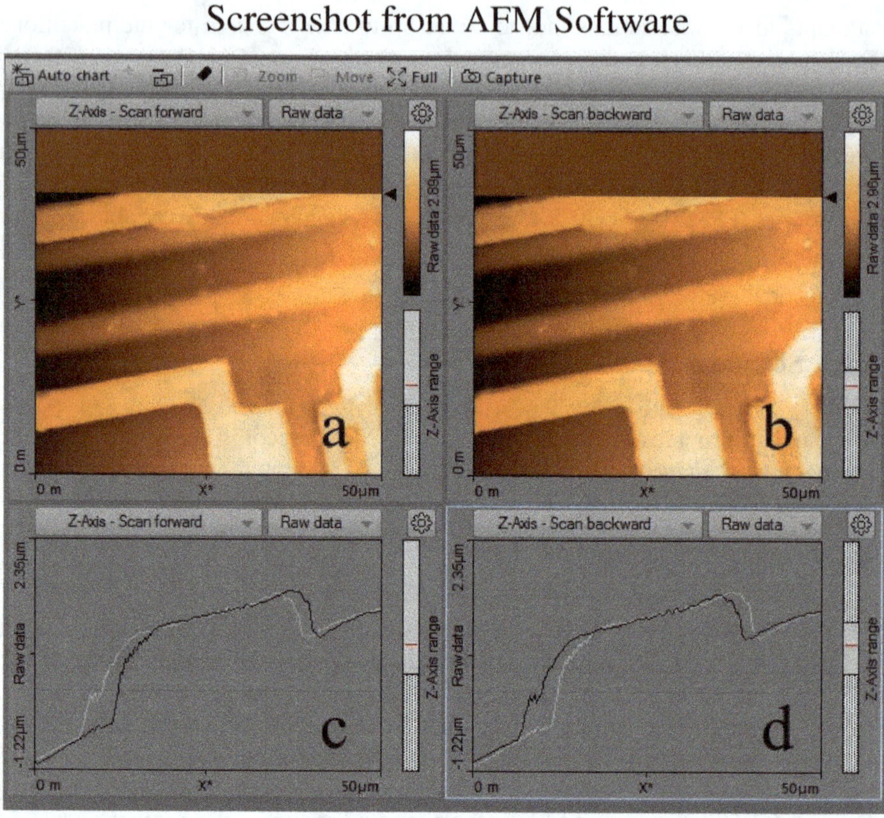

Fig. 1.4 A screenshot of an image being acquired: (a) and (b) are forward and backward images, while (c) and (d) are the respective cross-sections of the fast-scan line that was just acquired. The sample is an integrated circuit. The sample's tilt and the hysteresis between the forward and backward scans are apparent.

In **Instrument Lab 2**, you become more familiar with the user interface.

1.2.3.2 Image processing

Image data are stored as a function of either the vertical deflection d or vertical scanner height z as a function of horizontal x, y position. (Imagine an $x - y - z$ coordinate system on the surface of the sample.) The nominal z-position can be taken from the signal sent to the scanner by the feedback circuit (Sect. 2.1). Because some scanners respond nonlinearly to the signal sent to them, some manufacturers provide a means of monitoring the actual scanner position in addition to the signal sent to them. In the following, some nonlinearity is assumed.

Figure 1.5 represents typical image processing and analysis. In Fig. 1.5a, $z(x, y)$ data of a nominally flat substrate with small rectangular features are shown. Subtracting a plane from the image and looking down the y-axis results in Fig. 1.5b, where the thin dotted horizontal line is to highlight that the edges of the image are higher than the rectangular features, such that the color scale (vertical gradient bar) would be dominated by the remaining curvature of the substrate. A parabolic fit to the image and its subsequent subtraction is required to flatten the substrate, as in Fig. 1.5c.

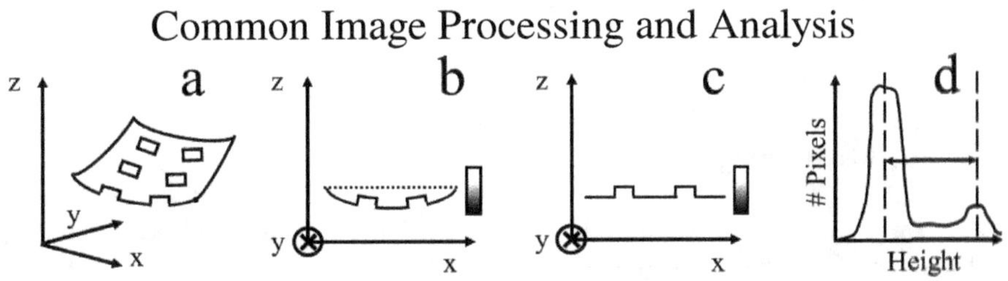

Fig. 1.5 Schematic representation of common image processing and analysis. (a) Raw data of small rectangular features on an apparently curved substrate; (b) After subtraction of a plane fit to the image; (c) after subtraction of a parabolic fit to the image; (d) a histogram of the data in (c), where the large and small peaks represent the heights of the substrate and tops of the rectangles, respectively.

If the height of the rectangular features is important to know, the researcher can make a cross-sectional measurement, starting with data as in Fig. 1.5c. But if the feature height varies slightly from feature to feature, as is often the case, then the researcher is faced with making many individual measurements and taking the average. Alternatively, the investigator can make a histogram, as in Fig. 1.5d, where the number of pixels at each height is plotted, much like a grade-distribution curve after an exam. The taller peak at the lower height reflects the height of the substrate and the smaller peak at the greater height the height of the features. Then a reliable value for the height of the features above the substrate is the distance between the peaks, the distance between the vertical dashed lines.

★ In **Computer Lab 1**, you learn how to process your raw AFM images. ★

This concludes the introductory chapter to AFM. Below you will find learning objectives and problems, the answers to which are in the back of the book. In the next chapter, some challenges associated with acquiring good images are discussed.

1.3 Learning objectives, further reading, and problems

! Learning objectives for Chapter 1

From the material in Chapter 1, the reader should now be able to:

1.1 Give examples of the size scale of AFM.
1.2 Distinguish between "top-down" and "bottom-up."

1.3 Demonstrate the operational principle of AFM.
1.4 List the disadvantages and advantages of AFM.
1.5 Describe a typical probe's size and composition.
1.6 Understand how an AFM collects $d(x, y)$ and $z(x, y)$ data.
1.7 Relate common AFM-related acronyms to their function.
1.8 Be familiar with basic image-processing options.

Further Reading

✓**History.** Binnig, G., Quate, C. F., & Gerber, C. (1986). Atomic force microscope. Physical review letters, 56(9), 930.

✓**History.** Meyer, G., & Amer, N. M. (1990). Simultaneous measurement of lateral and normal forces with an optical-beam-deflection atomic force microscope. Applied physics letters, 57(20), 2089–2091.

✓**Tutorial.** Stevens, S., Sutherland, L. M., & Krajcik, J. S. (2009). The big ideas of nanoscale science and engineering. National Science Teachers Association.

✓**Book.** Murty, B. S., Shankar, P., Raj, B., Rath, B. B., & Murday, J. (2013). Textbook of nanoscience and nanotechnology. Springer Science & Business Media.

? Problems for Chapter 1

1.1 Look online for an AFM image that interests you. Send the link to your instructor and be ready to describe it in class in a few sentences.

1.2 A fun run is a mile long. How far is this in (a) μm, (b) nm, (c) pm?

1.3 If a cantilever has a spring constant of 0.18 N/m, and it is deflected by 10 nm, how much force is applied to the sample? Answer in nN.

1.4 What is the magnitude of the pressure between a coffee cup (mass ~ 200 g) of diameter 10 cm and a desk. Use g = 10 m/s^2. Write the result in the correct SI unit for pressure.

1.5 If the tip of the cantilever in P.1.3 has a contact area with a sample of 10 nm^2, what is the pressure between the tip and sample? Use units of MPa.

1.6 If a cantilever has a spring constant of 45 N/m, it is bent backward 100 nm, and the contact area with the sample is 15 nm^2, what is the pressure between the tip and sample? Use units of GPa.

1.7 (a) How does the pressure in P. 1.6 compare with the yield strength of steel? (Look online for steel's yield strength. There are many different types of steel, so a typical value is fine.) (b) Do you think your tip and sample will remain unchanged under these conditions?

1.8 Sketch a schematic diagram of how an AFM works.

1.9 Distinguish between constant-height and constant-force modes.

1.10 Why would you use one mode over the other?

1.11 What are some advantages and disadvantages of AFM and SEM?

1.12 What are some commonalities of AFM and SEM?

1.13 Lab question. If you have acquired your first AFM images of a sample with regular features that was provided to you by your instructor, what is the spacing of the features? What is their height?

Chapter 2
Difficulties of imaging

Abstract Ironically, while the basic principle of AFM is simple to explain (Sect. 1.1.1), it is challenging to acquire high quality data. The most important element in acquiring good data is the feedback circuit; this chapter begins with an exploration of a typical feedback circuit. Recognizing and reducing image artifacts, some of which are related to the settings of the feedback circuit, is the subsequent topic. Since an AFM can be perturbed by electronic, mechanical, acoustical noise, by dust, and by variations in temperature, noise analysis is therefore introduced. Practical advice for situating and protecting an AFM from noise sources is then provided, in part based upon noise analysis. The chapter ends with an introduction to Fourier threshold and frequency filtering, both of which can help improve noisy images.

2.1 The feedback circuit

Why would you want to use a feedback circuit in your AFM? It is hard to insert a sample into the AFM that lies perfectly horizontally, considering that AFM can measure nanometer or Angström-level changes in height over microns laterally. If the scanner stays at the same height, then the cantilever will bend more over higher parts of the sample than over lower parts. The tip could even lose contact over lower parts of the sample.

Cantilever deflection is proportional to force. Remembering Hooke's Law, force is the negative of the *spring constant* (the spring's stiffness) times the deflection, or $\mathbf{F} = -k\mathbf{d}$. (Both force and deflection are vectors and are therefore shown in **bold** font. They point in opposite directions; hence the minus sign. Often only the magnitudes are important, so the equation is sometimes written as $F = kd$.) Thus, the force between the tip and sample can be vastly different across a sample if the scanner stays at a constant height. This problem should be avoided unless the sample is very flat, horizontal, and high scan speeds are required for the investigation.

Hence, AFMs incorporate a feedback circuit to try to keep the cantilever's deflection constant, which in turn means that the force between the tip and sample nominally remains constant. Then different force levels within an image can be avoided, in particular high forces that might damage your sample or tip. Image interpretation is more straight-forward if the force remains constant during data acquisition.

How does the feedback circuit work? As shown in Fig. 2.1, which is an updated version of Fig. 1.3, the feedback circuit takes a set point and a signal from the photodiode as inputs. The set point is the desired value of the parameter that is to be controlled. In the case of constant-force contact mode, often the default acquisition mode, this is the voltage from the photodiode representing the deflection of the cantilever; the conversion factor between volts and deflection is called the *sensitivity*. Sensitivity can be expressed in either volts per distance or its inverse, distance per volt. In the case of most dynamic-force modes, a lock-in amplifier, which measures the amplitude and phase of a repeating input, provides a signal against which the set point is compared. The comparison is done by subtracting the set point from the parameter that is to be controlled, resulting in the *error signal*. The error signal is zero if the input equals the set point.

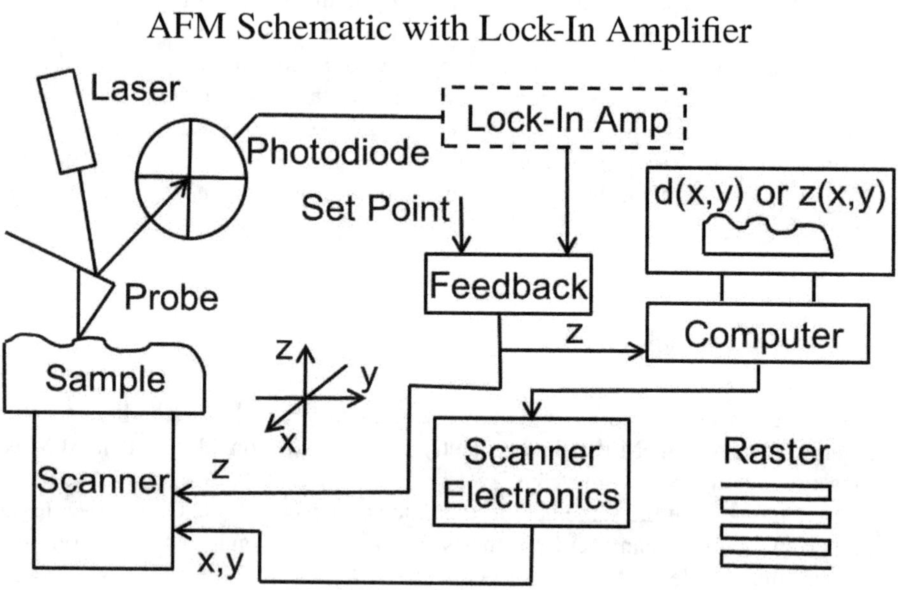

Fig. 2.1 An updated version of Fig. 1.3, now including a lock-in amplifier. When selecting a dynamic mode of operation, the lock-in amplifier provides the signal against which the set point is compared. It is bypassed for constant-force contact mode, where the deflection of the cantilever is compared to the set point.

While a detailed description of dynamic AFM modes is delayed until Sect. 3.4, a brief overview is given here, such that the reader will be able to understand how the feedback circuit can control both constant-force and dynamic modes. In a dynamic mode, the cantilever is oscillated at its natural frequency, the frequency at which it naturally oscillates. You might have played with a spring and vibrated it at different speeds and noticed how certain speeds generate large up-and-down motion of the spring. This behavior is the phenomenon of resonance. You can put a mass at the end of the spring and see the same kind of behavior. For a mass at the end of a light spring, the radial frequency at which this large motion occurs equals the square root of the spring constant over the mass, $\omega = \sqrt{k/m}$. For a cantilever, because its mass is distributed along the cantilever's length instead of being concentrated at the end, an effective mass is used. And please remember how radial frequency ω, measured in radians per second, is related to regular frequency, ν, measured in oscillations per second or Hertz, by $\omega = 2\pi\nu$; there are 2π radians per oscillation.

In a dynamic mode, the user will choose a frequency at which to vibrate the cantilever. The choice is usually a frequency close to its natural frequency. The lock-in amplifier "locks in" and determines the amplitude of oscillation at that frequency and provides that value to the feedback circuit. In a dynamic mode, it is the amplitude of oscillation that is compared to the set point in the feedback circuit, rather than the deflection of the cantilever, as for constant-force contact mode.

2.1.1 PID feedback

Within the box labeled "Feedback" in Fig. 2.1 is a circuit that modifies the error signal by a Proportional, Integral, or Differential calculation; see Fig. 2.2a. These feedback circuits are therefore called PID feedback. The calculations are added and sent on to the computer and scanner as the z (or height) signal. The signal to the scanner is often passed through a high-voltage amplifier (not shown) first, if large scan ranges are desired.

There are many different approaches to feedback design in AFMs. Most AFMs have proportional and integral feedback; some also have differential. Some AFMs will allow the user to adjust the strength ("gain") of each of the PID components, others are set up for simpler use with just one user control. Still other AFMs are sophisticated enough that the user can opt for the system to choose its own feedback settings based on the previous scan line of data. While this last option is good to start with, the user can sometimes be fooled into thinking that the collected data are a good representation of the surface. It is wise to verify any automated or default feedback settings with the simple tests described below.

In **Instrument Lab 3**, you examine the effects of PID feedback.

2.1.2 Feedback optimization

As explained above, the feedback circuit tells the z scanner to move up and down in response to the sample's topography. Users generally want the feedback circuit to

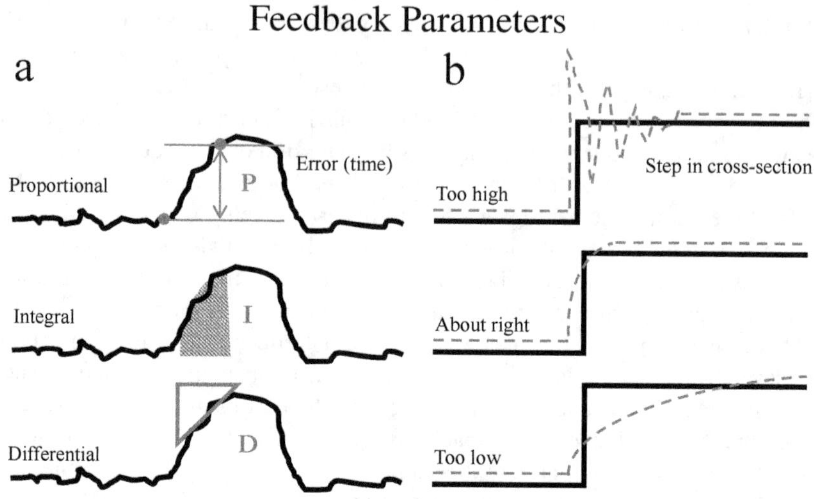

Fig. 2.2 (a) Representations of Proportional (top), Integral (middle), and Differential (bottom) calculations (gray) on the error signal as a function of time (solid black lines). Based on settings by the user, they are added together in different strengths ("gains"), which provide the height (or z) signal to create the image in the computer and adjust the height of the scanner. (b) Schematics of a cross-section of a topographic step (solid black lines), with the dashed gray lines representing the path of the tip. At the top, the gains are too high, in the middle they are about right, and on the bottom they are too low.

respond as quickly as possible such that image acquisition time is minimized. Indeed, one reason for the shift between the forward and backward scans ("hysteresis") in Figs. 1.4c&d could be feedback gains that are too low. (Another reason could be the effect of friction on a long cantilever; see Sect. 3.3.)

The natural temptation is to robustly raise the feedback gains. Unfortunately however, feedback gains can be set too high, resulting in the circuit being too responsive and oscillating, resulting in noisy images. Figure 2.2b is a schematic showing a cross-sectional step on a surface along with representations of the trace of a scanner subject to feedback gains being too high, about right, and too low. Note that even the "about right" trace is not perfect. Indeed, no AFM is capable of perfectly tracing the contours of a surface; an AFM will not move in the z-direction unless the feedback circuit receives a non-zero error signal. Hence, please remember that an AFM image is a representation of a surface's topography, and not a perfect one. Your job as an investigator is to minimize the error by optimizing the gains and provide the best possible representation of a sample's surface as an AFM topography image.

The gains are optimized in two principle ways. *First*, on the part of your sample with the tallest features that are closest together, raise the gain until oscillation sets in, then slowly lower the gain until the oscillation goes away. If your AFM has more than one gain setting (PI or PID, Sect. 2.1.1), test the different gains one-by-one to see which one has the greatest influence. Be aware that the sharpness of the sample topography and the other scan parameters will influence the optimal settings and the sensitivity of the settings to a good set of values. Feedback being one of the most important aspects of an AFM—or more broadly, any SPM—rather than mentioning here which gain setting has the most influence, readers are strongly encouraged to try it themselves, either informally or as part of a course-related laboratory, so that they are more likely to understand and remember; see P. 2.10.

The *second* way is more systematic and time consuming, but it is worth doing as you become familiar with your AFM. It is called "bracketing." If you have studied photography, you might have learned that photographs of the same scene can be taken with different exposures. A photographer might start with exposures that are likely to be too low, then step-by-step take pictures at higher exposures until it is fairly certain that the exposure is too high. Then the photographer can choose which photograph has the best exposure for the situation.

Similarly, you can systematically change the gains (and some other scan parameters, such as set-point force, the scan rate, or the number of pixels per line) during an image to see which settings result in the best images. Some AFMs will allow you to disable the slow scan, such that the cantilever scans back-and-forth on nominally one line. The associated image will then be a "waterfall plot," where the slow-scan direction represents the time of data acquisition. Most AFMs will allow you to simultaneously acquire error information along with a topography image. In constant-force contact mode, the error is usually displayed by monitoring the AC (real-time) components of the "deflection" image and in dynamic modes, the "amplitude" image. You can better see the result of small changes in gains and other scan parameters by examining the error images. As mentioned above, in a perfect world the error would be zero. Minimizing the error will result in the best possible topography image. Sadly, even with careful image optimization, there can still be artifacts in your data. The next section guides you on what to do.

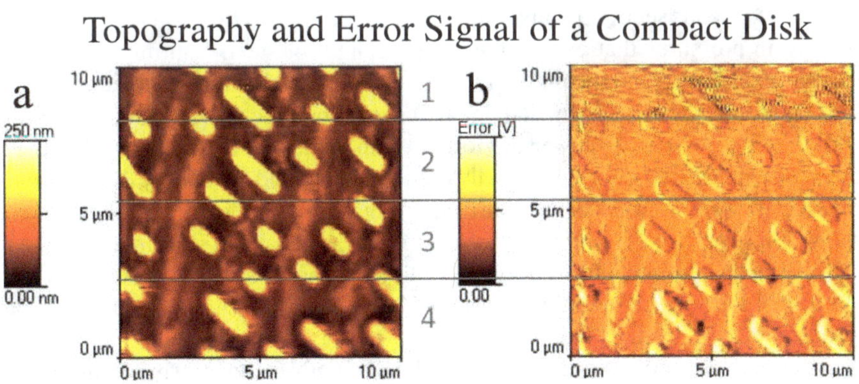

Fig. 2.3 AFM images of a pressed compact disk: (a) topography and (b) error signal. The data were bracketed so as to observe the effect of gains. The gray horizontal lines divide the images into four numbered regions for Problem 2.3.

AFM Lesson 6 at youtube.com\AtomicForceMicro
AFM Lesson 6 reviews the material in this section on feedback.

2.2 Artifacts and their remedies

You probably have an idea of what AFM images of your samples might look like. If your first images do not support your idea, there might be one or more things wrong in the AFM settings that is or are influencing the image quality, in other words, one or more artifacts. This section covers typical artifacts and what to do about them.

2.2.1 Typical artifacts

The first step in diagnosing and correcting for artifacts is that you know what they are! The most common ones are described here, along with suggested corrections.

2.2.1.1 Feedback gain too high or too low

Feedback gains being too high or too low is the most common hurdle in AFM imaging, deserving a section of its own, above. In particular, read Sect. 2.1.2 on feedback optimization for two ways to optimize the feedback settings.

2.2.1.2 Set-point force too high or too low

If the set-point force is too high, you might damage the tip or even the sample. While nano-Newton loads seem very small indeed, they are applied over square nanometers, resulting in pressures that exceed the yield strength of steel. You might not easily observe small changes to the tip. To look for damage to the sample, increase the scan range and lower the force. Any damage will show up in the middle of the new scan.

Yet you do not want to lower the force so much that the cantilever will not snap onto a lower lever of the sample. Ideally the scanner should move only the half-width of the tip in the time it takes the cantilever to bring the tip down to the lower level. You will observe this “parachuting" phenomenon as streaks in your image as the tip moves from an upper to a lower level. The direction of the streaks will change as the direction of the scan changes. An example is in Region 4 of Fig. 2.3, which can also be caused by gain settings that are too low.

As a starting set-point force, you might choose 50 nN for contact mode and 80% of the oscillation amplitude for a dynamic mode, then make adjustments as necessary. If it is important to preserve your tip and sample, use as low a force as possible and slow down the scan speed such that the feedback circuit has more time to move the scanner in the z direction.

2.2.1.3 Scan or pixel speed too fast for feedback loop

The electronic and scanning components of feedback circuits have their limits. They need time to stabilize. Similarly, cantilevers with low natural frequencies or low damping coefficients (Sect. 4.1.1) need more time to settle to a new value. The entire feedback loop cannot typically respond faster than a millisecond, although some fast-scanning (that is, expensive) AFMs might trim these response times into the microsecond range.

If the scan or pixel speed is such that the error signal can never settle down to zero, your images will not be a good representation of the surface topography. The scan speed is twice the scan range times the scan frequency; the scan frequency is the number of times the scanner goes back and forth in one second, measured in Hertz (Hz). Similarly, the pixel speed is how many pixels per second are acquired; this is the scan speed times the number of pixels in the scan range or twice the number of pixels times the scan frequency. Note that both are dependent on the scan frequency.

For example, you might start imaging with 25 μm at 1 Hz, which is 50 μm/s. If you set the number of pixels to 256 per 25 μm line, the pixel speed is about 500 pixels/s, allowing just 2 ms acquisition time per pixel. If the image quality is poor, try slowing the scan frequency for a few lines to see if image quality improves. If so, keep the lower frequency. If not, return to the higher frequency so as to not delay your total image acquisition time, and then look for other reasons why your images are bad (keep reading!). Yet remember that long acquisition times, say, greater than fifteen minutes, can increase the likelihood of thermal drift becoming a problem.

2.2.1.4 Thermal drift

A micron might not seem very big to us humans, but if you start a 5-μm image that takes an hour, it is possible that by the time the image finishes, the tip could be offset from what you wanted to image by a noticeable amount.

Most labs are subject to temperature variations during the day. Reasons could be from sun exposure, preset building cycles, equipment being turned on and off, or people gathering. Most materials have coefficients of thermal expansion on the order of $10–100 \times 10^{-6}$ m/m°C, so a one-degree Celsius change in temperature of a one-meter object results in a micron change in length. Most AFMs are exposed to fluctuations in room temperature, so they change slightly in shape, and frequently the relative position of the tip and sample change as a consequence. Image acquisition times of fifteen minutes or less are hence recommended. The image acquisition time is determined by the scan frequency, the number of lines in your image, and the number of pixels per line.

2.2.1.5 Number of lines or pixels too high or too low

The AFM samples and stores the image data numerous times per image line. The user controls how many times by inputting the desired number of pixels per line and the number of lines. A small square image intended as one of several in a published figure might comprise 128 pixels per line and 128 lines, for a total of 16,384 pixels. If each pixel takes 5 ms to acquire, then the total image acquisition time is about a minute and a half. More pixels mean longer total times. If rectangular images are acceptable, total image times can be shortened by restricting the number of lines, although the convention is to scan the same distance in the two horizontal directions and present the image in a square format. If an image is to be published in a larger format, perhaps two per figure in a publication, most investigators will use 256 pixels and 256 lines, increasing the total image acquisition time if all other parameters remain the same. Many users will initially use a relatively small number of pixels for a "pre-scan," then increase the pixel number when they find a position of interest on the sample.

Another consideration is how accurately the pixels track the features of the sample. Figure 2.4 demonstrates how too few pixels (top, light gray pixels and line) could completely miss the square-wave nature of the shown black cross-section at the bottom. Doubling the number of pixels over the same distance results in a sine wave (middle, gray pixels and line). Only when the pixel line density is quadrupled does the square wave become evident (black dots along the square wave). In this example, the pixel line density could be increased still more in order to accurately follow the vertical steps.

At some point as the pixel line density is increased, the accuracy of step replication will be limited by the shape of the tip, so it is wise to bear the tip-imaging phenomenon in mind, as described in Sect. 2.2.1.7. For now, let us consider what has been presented thus far.

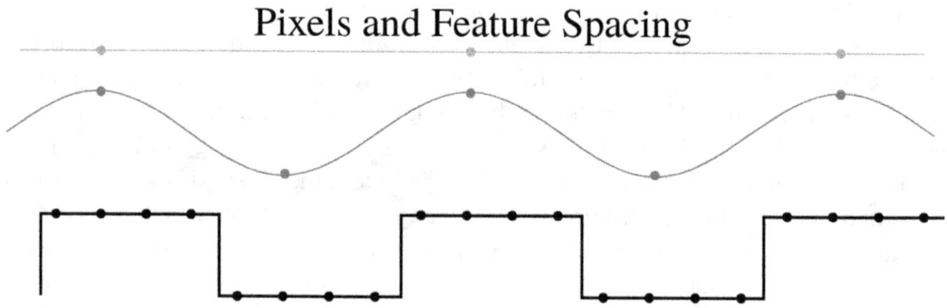

Fig. 2.4 Representation of the effect of the number of pixels relative to feature spacing, where the features come in a square-wave pattern (black line). Top, in light gray: the pixels (dots) are so few that an AFM image would be just a flat line. Middle, in gray: doubling the number of pixels over the same line length would lead to a sine wave in the AFM image. Bottom, black dots: quadrupling the number of pixels would finally start to reveal the square-wave pattern in an AFM image.

2.2.1.6 Compromises of imaging

Life is a set of compromises. So it is with AFM imaging. If you are still new to AFM, what you have read up to this point in this chapter might seem overwhelming. Table 2.1 is a summary of the scan parameters discussed so far and their influence on the quality of AFM images.

Table 2.1 The effects of extreme values of scan parameters on image quality.

Scan parameter	Too high	Too low
Feedback gain(s)	Noisy image	Poor tracking
Set-point force	Damage	Poor tracking
Scan or pixel speed (scan frequency)	Poor tracking	Thermal drift
Pixel spacing	Thermal drift	Poor resolution

In Sect. 2.1.2, you saw that too high feedback gains lead to noisy images and too low ones to poor tracking of the sample's topography. In Sect. 2.2.1.2, you read that too high a set-point force can damage the tip or sample (or both), and with too low of a set-point force the tip does not promptly follow the sample's contours. If the speed of the tip relative to the sample (or relative to the pixel spacing) is too fast, the feedback loop cannot keep up, and again poor tracking is the result (Sect. 2.2.1.3). Both scan speed and pixel acquisition rate ("pixel speed") are dependent on the scan frequency. Contrarily, if the scan or pixel speed is too slow, the image acquisition time might be long enough such that the sample drifts relative to the tip because of expansion or contraction due to temperature changes of the laboratory (Sect 2.2.1.4). And if your pixels are spaced more closely together than they need to be, image acquisition will be slow, and again thermal drift can be a problem. Yet as seen in Sect. 2.2.1.5, too large a pixel spacing can miss the features of interest.

And we are not done yet...there are more considerations for high quality images below.

2.2.1.7 Tip imaging

Does the tip image the sample? Or does the sample image the tip? Or do both happen?

The ideal case is that the tip is smaller than the features that need to be imaged, and the resulting image can be taken as a good representation of the sample surface. Sometimes, though, there are very sharp features on the sample that the tip must go over. In that case, the image that will be collected will be an inverted image of the tip, as sketched in Fig. 2.5a. This effect is called tip imaging.

Of course, every feature on the sample has a finite size. Imagine that the top of the spike in Fig. 2.5a has a radius of curvature of 10 nm, which is the same radius as many commercial AFM tips when they are new. Then 10 nm is imaging 10 nm, and the apparent width of the top of the spike will be 40 nm. (Fortunately, the detected heights are not affected.) In this situation, the tip and sample are imaging each other. The mathematical term for this process is *dilation* by the tip.

What can one do to determine if tip imaging is influencing the images? Take an initial image and note the relative positions of the features to each other and their orientation relative to the image. In Fig. 2.5b, three taller features form an isosceles triangle. Each feature is elongated towards the top of the image. Rotate the sample. If the features change both their placement and their orientation by 90°, the image is mostly free of tip imaging. Figure 2.5d shows how tip imaging can influence the rotated sample—the placement rotates, but the orientation of the features does not.

Bear in mind that changing the fast-scan direction by 90° degrees is not the same as rotating the sample by 90°. In the former case, the features' orientation and the tip's orientation remain the same. In the latter, they are different by 90°, allowing a determination if tip imaging contributes to the image.

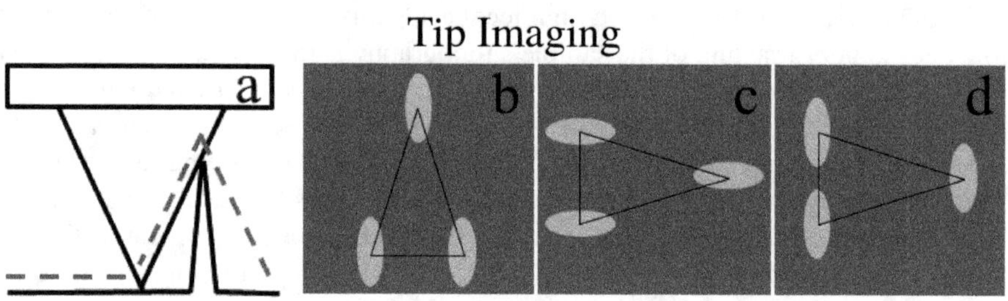

Fig. 2.5 Schematic illustrating tip imaging in AFM images. In (a), a cross-sectional view of a spike on a sample being traced out by a tip in contact mode is shown. The gray dashed lines indicate the motion of the tip; the image will appear as an inverted picture of the tip. In (b), three high features form an isosceles triangle in a sketch of an AFM image. A triangle is superimposed on the sketch to aid the eye, with the base of the triangle at the bottom and its long axis oriented vertically. In (c), the sample is rotated 90° clockwise. The base of the isosceles triangle is now on the left side of the image instead of at the bottom and its long axis is horizontal. The orientation of the high features rotated as well—the long axes are now horizontal instead of vertical. In (d), the sample is also rotated 90° clockwise, but the orientation of the features remains vertical, as if the detected feature shape was dominated by the shape of the tip, not the sample.

AFM Lesson 7 at `youtube.com\AtomicForceMicro`

AFM Lesson 7 briefly presents how to diagnose tip imaging.

● In **MacroLab 2**, you explore tip imaging. ●

2.2.1.8 Start-up artifact

Another artifact caused by the tip's close interaction with the surface occurs upon the beginning of a scan in contact mode. In contact, friction acts on the tip of the cantilever, and the tip and free end of the cantilever twist. (Refer back to Fig. 1.1 for a reminder.) At the end of a scan line, the top of the tip is twisted towards the direction of travel. As the scan direction is reversed, the scanner begins to move in the reverse direction and the tip twists into the reverse direction before the tip starts to slide with respect to the surface. Ideally, the resulting twisting and vertical motion of the free end of the cantilever would be completely decoupled, but they are linked; the twisting moves the reflected light spot slightly up and down on the photodiode in addition to right and left. Hence, the feedback circuit interprets the cantilever's motion as if topographic features were present. The start-up artifacts are seen on the beginning edges of the scans, as highlighted in Fig. 2.6, and they are of inverted contrast.

If contact-mode images are desired — perhaps for simplicity of use or interest in nanoscale friction — but the start-up artifacts are not wanted, most AFMs will allow an overscan. An overscan means that the image data are collected over a smaller range than the motion of the scanner. The scanning conditions of Fig. 2.6 were chosen to exaggerate the start-up effect: contact mode, flat sample, small scan range, and no overscan.

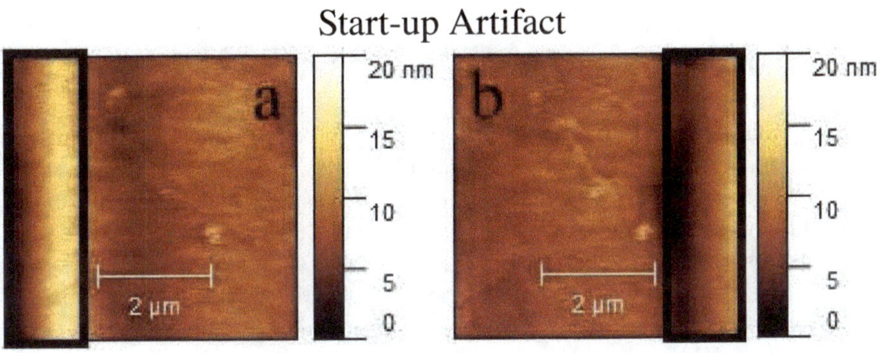

Fig. 2.6 The surface of graphite in contact mode with no overscan. Start-up artifacts inside the black highlight rectangles are evident on the left side of the forward scan in (a) and the right side of the backward scan in (b). They originate from friction on the tip and the resulting motion of the free end of the cantilever.

2.2.1.9 Scanner responding improperly

Although generally scanners work very well and can be controlled precisely, for example, as to enable the overscans described above, occasionally scanning fails. An example of poor *x-y* scanning is displayed in Fig. 2.7a. The sample was a recordable compact disk, for which the tracks are parallel on the AFM scale (although are circular on the human one!). Proper scanning of the same sample over a slightly larger scan range is presented in Fig. 2.7b, in which the parallel lines extend over this AFM's entire scan area. Keeping a known sample available that has regular features over the largest scan range that your AFM can manage is recommended, such that scanning can be checked occasionally.

As a side note, with respect to Sect. 2.2.1.5 on the number of pixels in an image, the images in Fig. 2.7 contain 128 × 128 pixels. If the images were reprinted any larger than in this small format, the reader would be aware of the pixelation.

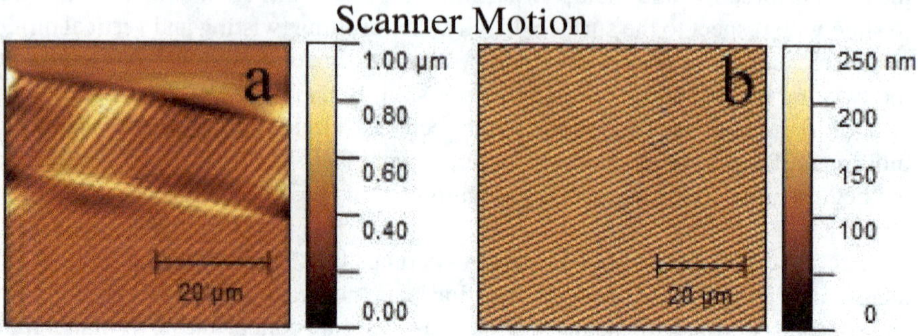

Fig. 2.7 Comparison of bad (a) and good (b) scanner motion of a recordable compact disk with nominally parallel tracks. In (a), the scanner's motion was blocked by an obstruction. Correct scanning is shown in (b).

2.2.1.10 Misleading image processing

Even if image acquisition is well done, images can be presented in such a way that they are misleading. A classic example involves background subtraction. If a scanner responds slightly imperfectly, that is, moving not in a horizontal $x-y$ plane, but with some curvature, then a perfectly flat substrate (such as a silicon wafer) will appear curved, as in the bowl shape in Fig. 1.5a. if the user uses only a planar background subtraction, Fig. 1.5b is the result, and the substrate still appears curved. Subtracting a parabola from the raw data yields the flat substrate in Fig. 1.5c.

A learner of AFM should be alert to what the image processing program is doing, such that a final image is a good representation of the sample. The most difficult aspect of image processing to understand is Fourier filtering, which is covered in Sect. 2.5.

Now that different artifacts have been discussed, the next section presents approaches to artifact detection.

2.2.2 Artifact detection

How does one quickly check to see if images are free of artifacts? They are probably free of artifacts if: (i) the features do not change when the scan is repeated or the scan direction is changed, (ii) the features scale when the scan range is changed, and (iii) the features' relative position and orientation stay the same when the sample is rotated. This third check is often skipped to save time, but is useful in the context of tip imaging (Sect. 2.2.1.7).

Sometimes initial images are particularly puzzling. To more systematically diagnose artifacts, you can follow this list of steps:

- Check number of pixels per line.
- Check scan frequency.
- Check pixel acquisition speed (= 2 • pixels per line • scan frequency).
- Compare forward and reverse images to check for hysteresis.
- Check set-point force.
- Adjust the gain(s).
- Change direction of fast-scan motion.
- Check for proper scaling of features by changing the scan range.
- Rotate the sample to check for tip imaging.
- Image a known sample to ensure proper scanning motion.

Once you achieve good images for your sample, make note of your optimal scan parameters; some AFMs will allow each user to save configurations. If you return to the same kind of sample in the future, you will save time in image acquisition if you begin with the same settings. However, keep in mind that samples with significantly different feature heights and spacing might require different parameters for good images.

If after following these steps, you still do not understand your images, it might be time to ask your AFM's manufacturer for advice. If you see the sample's expected features, but they are somewhat obscured by noise, then keep reading — the next section on noise analysis is for you. And once you have address any noise problems as best you can, you still have the last resource of using image processing to clean up your data. One approach to image processing, the most conceptually difficult one of Fourier filtering, is described in Sect. 2.5.

AFM Lesson 8 at `youtube.com\AtomicForceMicro`

AFM Lesson 8 summarizes image optimization and artifact detection.

2.3 Noise and perturbations

In addition to incorrect feedback settings and the wide range of artifacts in AFM, noise is a very important topic. After all, you are trying to measure distances with the precision of an Angstrom — the diameter of a hydrogen atom — which requires the noise level to be substantially less than an Angstrom. The analysis of noise helps determine if your AFM, or in general, your SPM, is adequately protected from mechanical, acoustical, and electromagnetic perturbations. Noise analysis is as well the first step in understanding an advanced method to improve images called Fourier filtering.

2.3.1 Noise analysis

With experience, a look at an oscilloscope trace of your microscope's error signal (Sect. 2.1) can quickly tell you if your data will be noisy or not. But it is often difficult to diagnose the cause of a noisy signal from the oscilloscope, which shows the error signal voltage as a function of time. It is much easier to understand the noise if it is presented as a function of frequency.

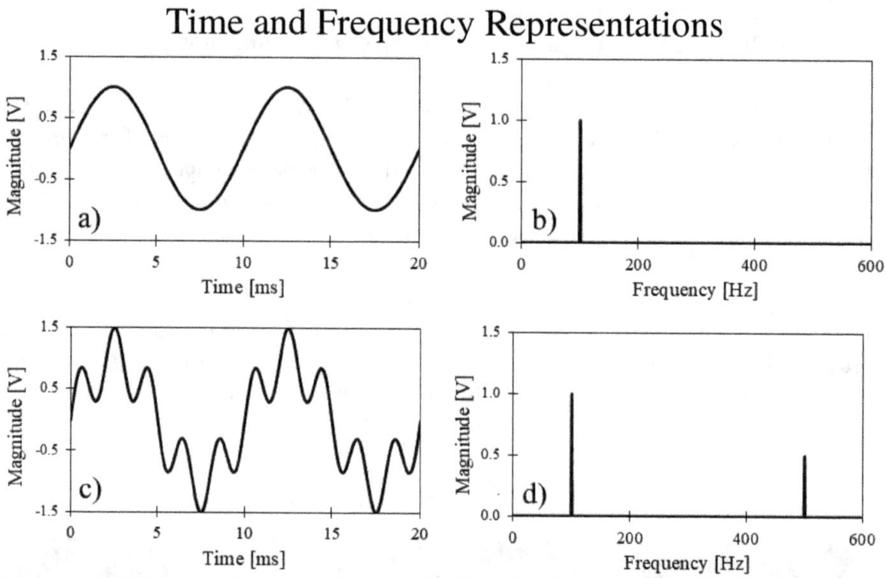

Fig. 2.8 Time (a, c) and frequency (b, d) representations of (a, b) a sine wave and (c, d) the sum of a 1 V, 100 Hz sine wave and a 0.5 V, 500 Hz sine wave. The characteristics of the summed wave are much easier to read from (d), the frequency representation, than (c), the time representation.

Take, for example, the sine wave shown as voltage as a function of time in Fig. 2.8a. It can be represented very simply as a single spike on a graph of voltage magnitude as a function of frequency, as in Fig. 2.8b. The slightly more complicated waveform shown in Fig. 2.8c is the sum of two sine waves of different frequencies and amplitudes. Whereas it takes some thinking to deduce exactly which frequencies and amplitudes form the sine wave in Fig. 2.8c, it is immediately apparent from its representation as a function of frequency, Fig. 2.8d. One can easily read from Fig. 2.8d that the waveform in Fig. 2.8c is composed of a 100 Hz sine wave of magnitude 1 V, and a 500 Hz sine wave of magnitude 0.5 V.

Even very complicated time signals are easier to understand when presented as a function of frequency. The process of changing from the "time domain" to the "frequency domain" is called a *Fourier transform* (after Jean Bapiste Joesph Fourier, 1768-1830). The process of changing from the frequency domain to the time domain is called the *inverse Fourier transform.* The algorithms commonly used in computer programs are the fast Fourier transform (FFT) and its partner the inverse fast Fourier transform. Many modern oscilloscopes come equipped with Fourier transform capability and more expensive instruments called spectrum analyzers are expressly designed for collecting such data. A common way to present time-domain SPM noise data after they have been Fourier transformed is called a power spectrum.

2.3.1.1 Power spectra

A spectrum is a form of data presentation in which the x-axis is frequency. Before we go further, we must define some terms so that you know what to look for when you see power spectra. Whereas a *magnitude spectrum* has the magnitude of the variable on the y-axis, the y-axis of a *power spectrum* shows the mean-square value of the variable. The compound adjective "mean-square" concerns how averages are traditionally taken for fluctuating signals, such as the sine wave in Fig. 2.8a. Its average would be zero. However, if the sine wave is squared, a nonzero answer for the average would be found. The mathematical definition of the average (the *mean value*) of an arbitrary function of time $f(t)$ is represented by $\langle f(t)\rangle$. Hence the mean value of $f(t)$ is

$$\langle f(t)\rangle = \frac{1}{T}\int_0^T f(t)\, dt. \tag{2.1}$$

where T is the length of the interval over which the average is taken. For example, the mean value of one period of a sine wave with no offset is zero. The *mean-square value* of $f(t)$ is

$$\left\langle f(t)^2\right\rangle = \frac{1}{T}\int_0^T f(t)^2\, dt. \tag{2.2}$$

In the case of the square of the sine function, for one period of the sine wave, the mean-square voltage becomes

$$\left\langle \sin^2 \omega t\right\rangle = \frac{\omega}{2\pi}\int_0^{2\pi/\omega} \sin^2 \omega t\, dt = \frac{1}{2}, \tag{2.3}$$

with ω the radial frequency and T the period ($\omega = 2\pi\nu = 2\pi/T$). Be careful of this factor of two. The mean-square amplitude is half the square amplitude. Yet another often-heard term, the *root-mean-square (rms) amplitude*, is the square root of the mean-square amplitude; in this case, $1/\sqrt{2} = 0.707$.

Power is defined as the change in energy per unit time, measured in Joules per second, or Watts. At first sight, the units that are frequently used in power spectra,

$\langle V^2 \rangle$, the mean-square value of the voltage, do not correspond to power. What has been left unsaid is that the input resistance of most oscilloscopes and spectrum analyzers is 50 Ω. The y-axis should more properly be given as $\langle V^2/R \rangle$, which does have units of Watts.

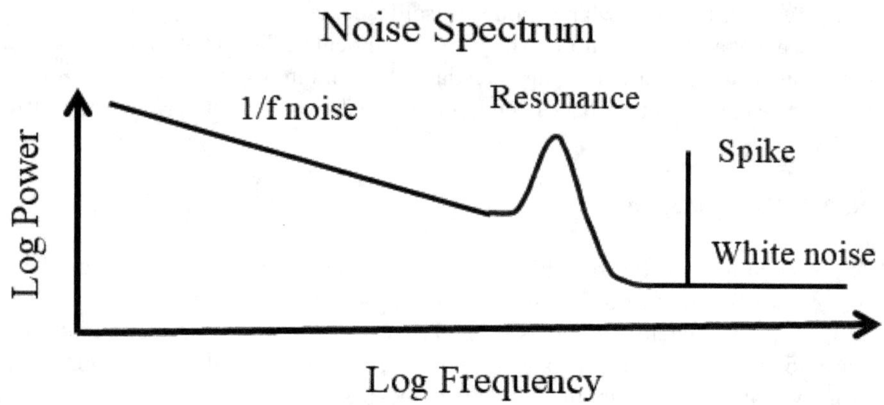

Fig. 2.9 Sketch of a typical noise power spectrum on in a log-log plot, showing the principle kinds of noise: 1/f noise, resonances, spikes, and white noise.

2.3.1.2 Types of noise

A number of different types of patterns can be discerned when noise is displayed in the form of a power spectrum. Typical types of noise are sketched in Fig. 2.9, where the logarithm to the base ten of the mean-square voltage (the power) is plotted as the logarithm to the base ten of frequency. When a large range of values must be displayed on the same plot, logarithmic scales are often used. These scales are useful in another way too. The slope of a log-log plot gives the mathematical power of the function. For example, in the first type of noise that we identify, *1/f noise* (pronounced "one over eff"), the slope is minus one. The power of 1/f noise is minus one, i.e. $1/f = f^{-1}$. The origin of 1/f noise was long mysterious; recently researchers believe that its source is many different kinds of defects at semiconductor interfaces within the AFM's electronics.

Resonances can be found in noise spectra. These patterns, wide peaks in noise spectra, are due to mechanical or electronic oscillators, which we will study in Chap. 7.

Perturbations that generate a discrete peak in a noise spectrum are variously called discrete or line noise or *noise spikes*. Often noise spikes originate by means of electromagnetic pickup from power lines, in which case they are usually at the power line frequency or a multiple thereof. They can also be caused by electromagnetic devices that radiate, such as monitors and the digital clocks found in computers.

Finally, *white noise*, as opposed to 1/f noise, has no frequency dependence and therefore appears as a horizontal line in a noise spectrum. White noise comes from the random motion of electrons within a circuit. Some amount of white noise is therefore unavoidable. It usually constitutes the lowest lying contribution or "floor" to a noise spectrum.

If your SPM signal is noisy, conduct noise analysis by making a power spectrum and observing the patterns in the spectrum as described above. The next step is to identify the probable causes of the noise. Large 1/f noise would indicate that your preamplifier is poorly designed. A spike at 50 or 60 Hz is a good indication that your microscope is picking up a signal from the power line. A high noise level at frequencies in the range of 10 to 1000 Hz likely means that building vibrations are propagating to the sensitive part of your instrument. Frequencies in the range of 100 Hz to 20 kHz correspond to audible signals, and a high noise level in this frequency range might mean that your instrument requires better acoustical shielding. After an attempt to rectify the problem, take another noise spectrum, then see if and how it has changed.

★ In **Computer Lab 2**, you explore feedback and noise. ★

This ends the description of noise detection. It's time to address noise abatement.

2.4 Situating and protecting your SPM

AFM Lesson 9 at `youtube.com\AtomicForceMicro`

AFM Lesson 9 discusses the material in this section on perturbations.

SPMs are designed so as not to be perturbed by their surroundings, but the noise specifications are demanding and hard to achieve. Constructed with stiff materials and designed to be as small as possible, the frequencies at which the mechanical parts of an AFM or SPM naturally vibrate are usually in the kilohertz range or higher. Building vibrations, in contrast, are usually lower in frequency (10-100 Hz). Any low-frequency vibration that reaches a high-frequency SPM does not cause much movement. To reduce the risk of mechanical vibrations even further, SPMs are best situated in a basement in a room off of a quiet corridor away from pumps, generators, and strong flows of air. Ideally, they should be on a vibrationally isolated floor. If they must be on an upper level of the building, they should be close to a major vertical support of the building and not in the center of a room. To further protect them from building vibrations, they should be placed on an air table (~1 Hz) or suspended from very low-frequency springs (~1 Hz). Even so, many students will prefer working at night, when mechanical vibrations are usually lower due to reduced activities in the building or neighborhood.

If you are looking for a location for an SPM, a quick test is to take your filled coffee cup and a flashlight with you on a building tour. Sprinkle a few small objects, e.g., bits of plastic, to float on the surface of the liquid, be it coffee or another fluid. Turn off the lights in the prospective laboratory room, and place your cup on the floor. After the fluid stops sloshing in the cup, shine the flashlight at a glancing angle at the surface of the fluid. Look for ripples in the menisci around the floating objects. If you cannot see any, the room might be good for an SPM. If you do see ripples, it is unlikely that you would ever obtain good images if you were to use the room as a probe microscopy laboratory.

As an example of a room that is not good for SPM while a nearby compressor is running, examine Fig. 2.10. There are two noise spectra, one with the compressor off and one with the compressor on. For the data acquisition from the accelerometer that was used, the noise power is reported as acceleration squared as a function of frequency in Hertz. Section 2.5.1 explains how if one adds all of the y-axis values together and takes the square root, the root-mean-square acceleration is found. The rms accelerations for the 1 Hz to 10 kHz frequency range for the two cases are 582 and 4510 $\mu m/s^2$ for the compressor being off and on, respectively. The compressor adds significant noise in the range of 100-1000 Hz. Ripples were not seen in the coffee cup in the former case, yet they were in the latter.

Even if the room passes the "coffee-cup test," further vibration isolation will likely be needed. Figure 2.11 demonstrates how an air table reduces noise. The noise spectra were taken when there was no pressure in the air table with the table-top resting upon its legs, and when the table was operating normally with the table-top suspended by pressurized air in the legs of the table. In the former case, building vibrations in the range of 10-100 Hz are easily transmitted through to the table. For the latter situation, the air table adds noise to the 1-10 Hz range, but eliminates much of the noise in the 10-100 Hz range. The properly functioning air table lowered the rms acceleration from 613 to 127 $\mu m/s^2$.

Comparing Figs. 2.10 and 2.11, one observes similarly shaped spectra and rms accelerations for the cases of the compressor off, with the accelerometer on the floor, and air table off, with the accelerometer on the air table. The two rooms in which the experiments were done are in the same basement; the room with the air tables is located farther away from the compressor.

SPMs also need protection from acoustical excitation. Sometimes one can observe a particular person's voice in the time trace of the error signal. That person's voice is just at the right frequency to excite a mechanical vibration in the instrument. If an acoustical isolation hood is not provided with your SPM, an inexpensive one can be constructed from a cardboard box lined with foam. Additionally, users of the lab and neighboring labs can be asked to speak and play music quietly and to close doors gently. A target value for acoustical noise in your laboratory is 30 dB, about the noise level of a whisper heard from a meter away from its source.

High 1/f noise can mean that an AFM's light source is noisy, or that its amplifiers were poorly designed. High electromagnetic noise can mean that an SPM is not properly grounded or shielded. If your instrument was purchased, this is unlikely to be true unless you must make some repairs and you leave some connections or

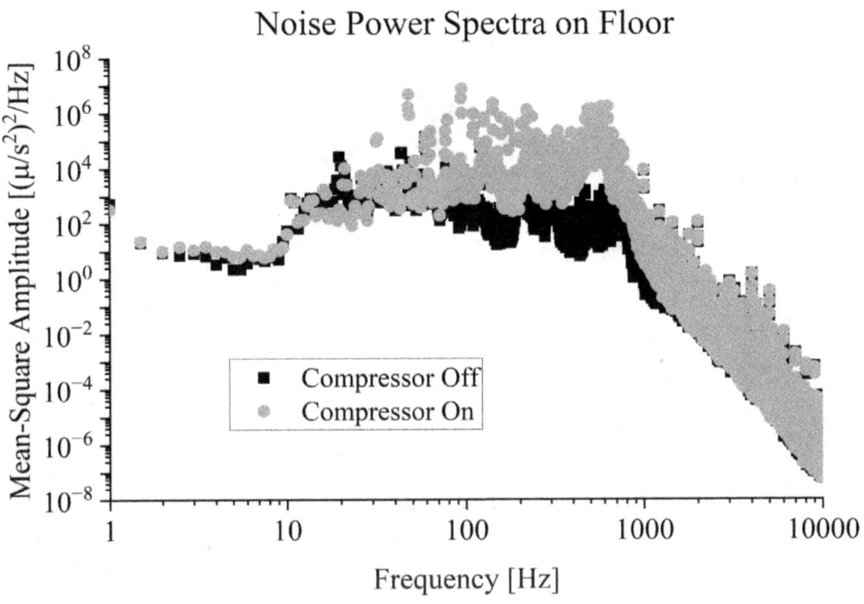

Fig. 2.10 Example noise power spectra from a floor in a room with an air compressor next door. The logarithmic y-axis scale is in square acceleration per Hertz, and the logarithmic x-axis is in frequency, also in Hertz. The floor vibrations are low when the compressor is off (black squares). However, the compressor makes the room unsuitable for probe microscopy data acquisition when the it is running (gray circles), causing an acceleration of 4510 $\mu m/s^2$ as compared to 582 $\mu m/s^2$ when it was off. With the compressor on, the room did not pass the "coffee-cup test."

covers undone. Another source of electrical noise is the building's power. A dedicated circuit, with only the instrument on it, is much less likely to permit glitches in the data when, for example, the air conditioning turns on, than if the circuit were shared with the air conditioning. Power conditioners can also help lower noise from the power lines; they are commercially available either independently or integrated into back-up power supplies. Some monitors emit high frequency radiation that is picked up by the SPM's electronics. These instruments can be removed from the immediate vicinity, or the electronics can be further shielded by grounded metal.

Example noise power spectra taken directly from an AFM are shown in Fig. 2.12. Here, the y-axis is the logarithm of voltage squared, and the x-axis is the logarithm of the frequency in kilohertz. A probe assembly was inserted into the AFM such that the laser light reflected from the chip, rather than a cantilever. The error signal was Fourier transformed to produce the spectra. With the laser switched off, the rms voltage fluctuations are 1.66 mV over the entire frequency range. With the laser switched on, the rms voltage rises by forty percent, to 2.32 mV. All four principal types of noise described in Sect. 2.3.1 can be found in the figure.

A final cause of noise, too low in frequency to be recorded in a normal power spectrum, is due to the changes in the size of mechanical parts if temperature cycles

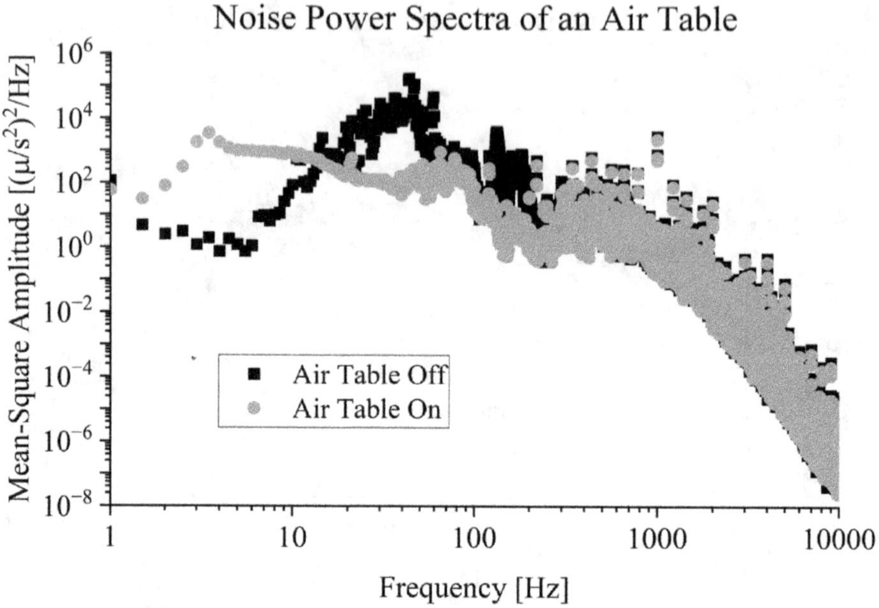

Fig. 2.11 Example noise power spectra from an air table that was first empty of air (black squares), then later filled with air (gray circles). The logarithmic y-axis scale is in square acceleration per Hz, and the logarithmic x-axis is in frequency, Hz. Although the air table adds noise in the 1-10 Hz range, it reduces the noise in the 10-100 Hz range, lowering the overall noise level, here measured in acceleration, from 613 to 127 $\mu m/s^2$.

exist in the lab. A typical meter-long object will expand a micron if its temperature changes by a degree Centigrade. A micron is very small on a human scale, but a thousand times bigger than a nanometer. Choose a lab without windows in which to situate your SPM in order to reduce the influence of daylight on room temperature. Try to maintain a steady temperature at all times.

Although not considered noise, dust is a perturbing factor in SPM experiments. Complete removal of particles from the atmosphere is not achieved even in the most expensive clean rooms, but dust levels can be minimized with a few simple precautions. Clean everything that goes into the lab. Place fine air filters over air inlets and outlets. Put clean-room adhesive mats at the entrances to the lab. Keep the doors closed and use weather stripping around them to reduce air circulation and prevent loud slams. Require users to remove their shoes or use clean-room booties over their shoes. Leave coats and food outside. A clean lab holds the further advantage that no one hesitates to sit on the floor when necessary. Finally, you should aspire to maintain constant humidity in the lab, as some experiments can be affected by it.

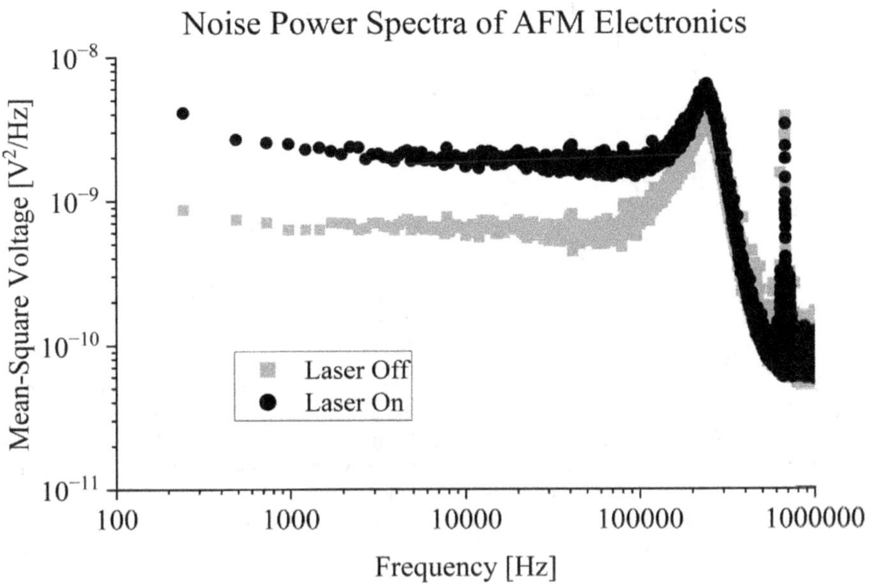

Fig. 2.12 Noise power spectra of an AFM with its laser off and on. The rms voltage fluctuations increase by forty percent when the laser is on. All four principal types of noise can be seen.

> **AFM Lesson 10 at youtube.com\AtomicForceMicro**
>
> AFM Lesson 10 discusses noise analysis and introduces Fourier filtering.

2.5 Fourier threshold and frequency filtering

Noise analysis is accomplished with power spectra, which are the result of Fourier transforming the error signal from your SPM. Even when the noise levels in your instrument are good, sometimes your images can be improved with post-acquisition processing based on Fourier filtering. After a brief summary of Fourier theory, the principles of Fourier filtering will be explained. An example of how Fourier filtering is used to improve SPM images is then demonstrated.

2.5.1 Fourier theory

The fundamental idea behind Fourier theory is that any periodic signal can be represented as a sum of a mean value together with cosine and sine waves of different

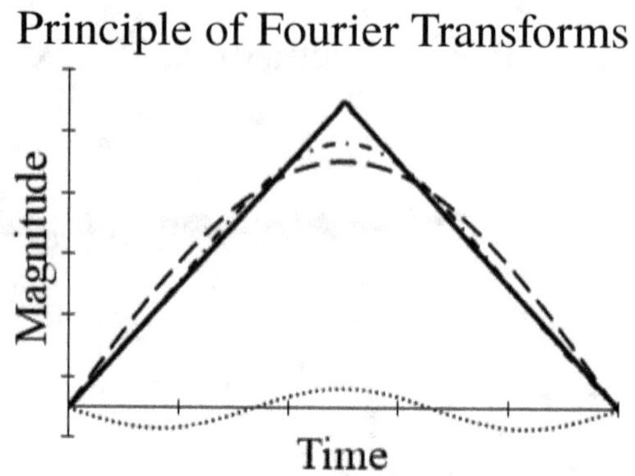

Fig. 2.13 Any periodic signal can be represented as a sum of an offset along with sine and cosine waves of different frequencies and amplitudes. Here, a half-period of a triangle wave is shown by the solid line. The dashed line is a sine wave of the same frequency, the dotted line is a sine wave of three times the triangle's frequency. When the two sine waves are added together, the result is the dot-dashed line, a better approximation to the triangle wave. By adding more sine waves of higher frequencies, a very good approximation of the triangle wave can be formed.

amplitudes and frequencies. Mathematically, this principle is expressed as the series function

$$f(t) = a_0 + \sum_{n=1}^{\infty} (a_n \cos n\omega t + b_n \sin n\omega t), \tag{2.4}$$

with a_0 the mean, a_n the amplitudes of the cosine functions, and b_n the amplitudes of the sine functions. As an example, look at Fig. 2.13. A half-period of a triangle wave (solid line) can be approximated by the sum of a sine wave of the same frequency (dashed line) and a sine wave of three times the frequency (dotted line). The sum appears as the dot-dashed line. Higher frequency waves can be added to construct a very good representation of the triangle wave.

Collectively, a_0, a_n, and b_n are known as the Fourier coefficients. They are the magnitudes of the frequencies in a magnitude spectrum, and they are found by integrating the time-domain data over an time interval of T, with

$$\begin{aligned} a_0 &= \frac{1}{T}\int_0^T f(t)\, dt, \\ a_n &= \frac{2}{T}\int_0^T f(t)\cos n\omega t\, dt, \\ b_n &= \frac{2}{T}\int_0^T f(t)\sin n\omega t\, dt. \end{aligned} \tag{2.5}$$

Multiplying Eq. 2.4 by $f(t)$, taking the integral in time from 0 to T, then dividing by T gives

$$\frac{1}{T}\int_0^T f(t)^2\,dt = a_0^2 + \frac{1}{2}\sum_{n=1}^{\infty}\left(a_n^2 + b_n^2\right). \tag{2.6}$$

Comparing the above with Eq. 2.2, one realizes that the left-hand side is the definition of mean-square value. Furthermore, from Eq. 2.3, $a_n^2/2 = \langle a_n^2 \rangle$. Hence,

$$\langle f(t)^2 \rangle = a_0^2 + \sum_{n=1}^{\infty}\left(\langle a_n^2 \rangle + \langle b_n^2 \rangle\right), \tag{2.7}$$

and the mean-square value of the function equals the sum of the mean-square values of the Fourier coefficients (the powers for the frequencies $n\omega$ in Eq. 2.4), plus the square of the mean. Effectively, by summing the y-values of a power spectrum, the mean-square value can be found. Equation 2.7 is known as Parseval's Theorem, which demonstrates the equivalence of finding the mean-square value in either the time (Eq. 2.2) or frequency (sum power spectrum y-values) domains. From this, by taking the square root, the root-mean-square value of the function is found, a commonly reported value in power spectra for noise analysis (Sect. 2.3.1.1).

The above discussion was limited to periodic functions. The same arguments can be generalized to nonperiodic functions, with the principles remaining the same.

2.5.2 Improving SPM images

Fourier transforms can be used to improve the quality of SPM images. If noise spikes exist in the data, they can be removed. If white noise interferes with image quality, it can be filtered out. Fourier filtering works especially well if there are periodic features in your images. As an example, we start by taking a triangle wave, adding white noise, then Fourier filtering it to recover a signal that is close to the original.

Figure 2.14a shows a 100 Hz triangle wave of amplitude 1 V. Its Fourier transform is in Fig. 2.14b. Adding white noise yields the waveform and spectrum in Figs. 2.14c and d, respectively. If a threshold corresponding to the horizontal dashed line in Fig. 2.14d is chosen, and thus any data below that value are set to zero, the spectrum becomes Fig. 2.14f. Inverse transforming again, the time domain appears in Fig. 2.14e. Comparing Figs. 2.14c and e, we see that some information has been lost. The peaks of the triangles are no longer sharp, yet the basic waveform is now free of noise.

Filtering SPM data is very similar in principle, but with two confusing differences. Firstly, data are acquired as a function of position, rather than time. Instead of time and frequency being the variable and its Fourier transform, "real-space" and "k-space" comprise the pair. K-space is the inverse of real space, just as frequency

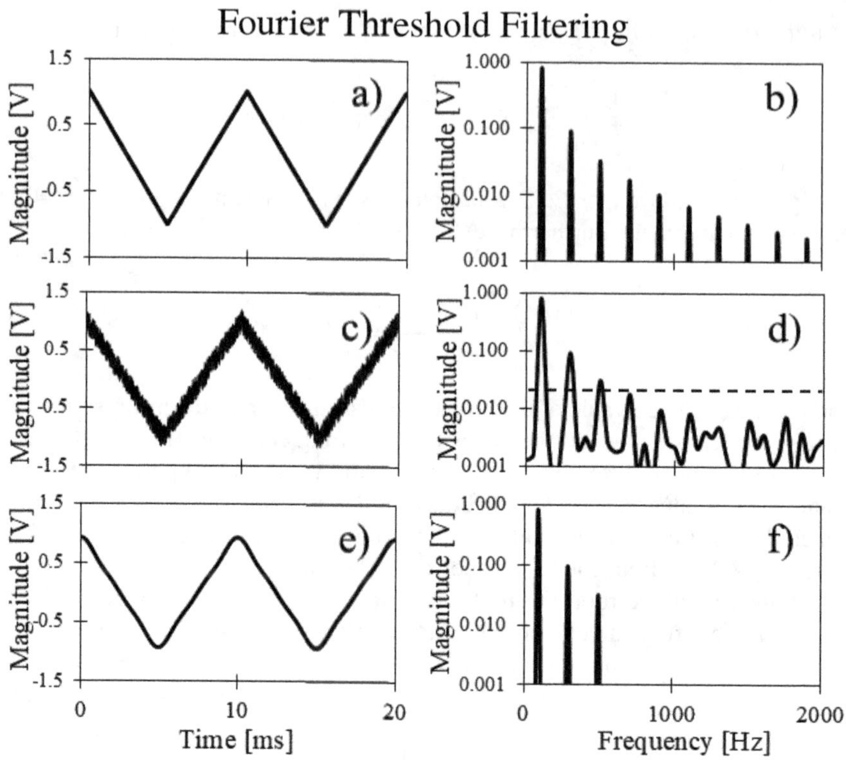

Fig. 2.14 Fourier threshold filtering. On the left are time-domain plots, on the right lie the corresponding frequency-domain graphs with their y-axes on log scales. In (a) and (b), a clean triangle wave with no noise is shown as a function of time and frequency, respectively. In (c) and (d), noise has been added to the triangle wave. A threshold corresponding to the horizontal dashed line in (d) is then applied, such that the peaks above the line are inverse-transformed into the time-domain graph in (e). The three peaks in (f) are those that comprise the signal in (e).

is the inverse of time; real space is measured in nanometers or microns, k-space is measured in inverse nanometers or inverse microns. Secondly, the data are now in two dimensions instead of one, and the spectrum is now displayed on a two-dimensional coordinate system that runs from negative μm to positive μm along each axis. The strength of the peaks is indicated by a gray or color scale running from dark (weak) to light (strong).

Two examples of two-dimensional Fourier transforms are presented in Fig. 2.15. Figure 2.15a is the original image of a square wave that runs in the x-direction, and Fig. 2.15b is its Fourier transform. The peaks lie along the x-axis, signifying that the sine and cosine waves that constitute the square wave are also oriented in the x-direction. In Fig. 2.15c, the square wave runs in the y-direction, and Fig. 2.15d shows its Fourier transform, with its peaks oriented along the y-axis.

An example of Fourier threshold filtering is shown in Fig. 2.16. Figure 2.16a is the original, streaky image, and Fig. 2.16b is its Fourier transform. The low-frequency

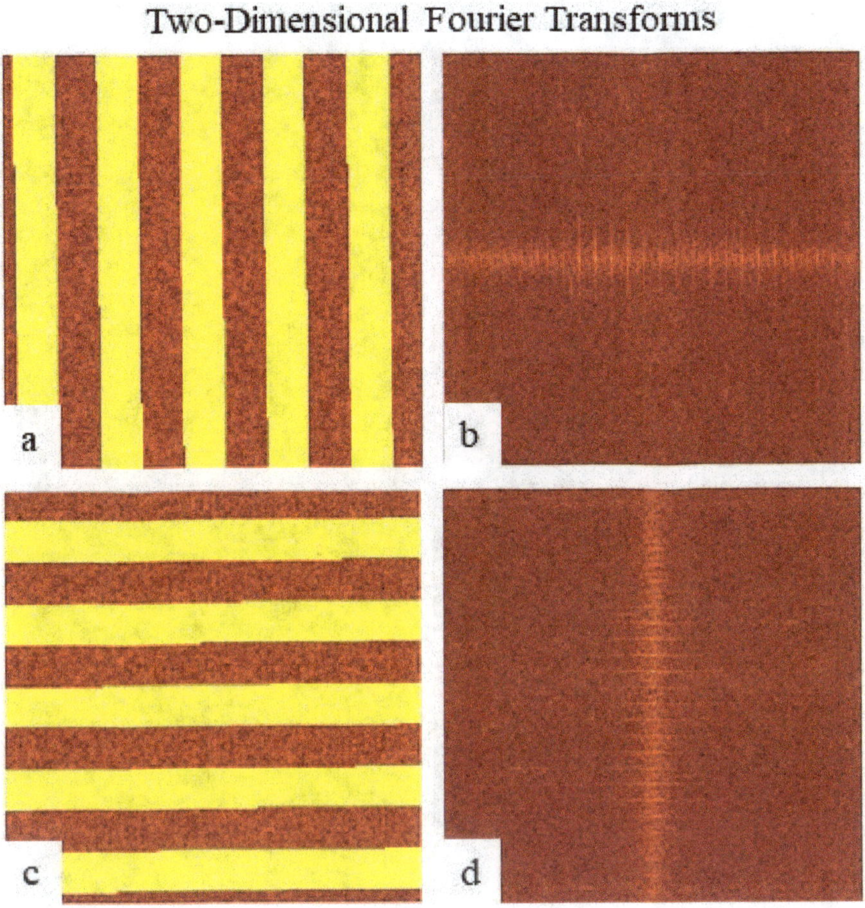

Fig. 2.15 Example two-dimensional Fourier transforms of square waves that run in one direction. The original images are (a) and (c). Their respective Fourier transforms are (b) and (d). The units of the axes in (b) and (d) are inverse microns, and the gray scale represents the height of the peaks, with white being the strongest value. The peaks in (b) and (d) lie along the axes along which the square waves in (a) and (c) are oriented.

peaks near the center are the brightest (highest amplitude); the spectrum has not yet been filtered. In Fig. 2.16a, a threshold of 78% of the maximum power has been applied. The strong peaks that are greater than 78% of the maximum power, now shown in gray and circled in Fig. 2.16c, are transformed into the filtered image seen in Fig. 2.16d, and the rest of the spectrum is ignored. The effect of the threshold filtering is to remove the horizontal streaks in the original image, yet preserve its periodic structure.

Figures 2.14 and 2.16 demonstrate Fourier threshold filtering. One can also filter images by keeping or deleting specific frequencies, known as Fourier frequency

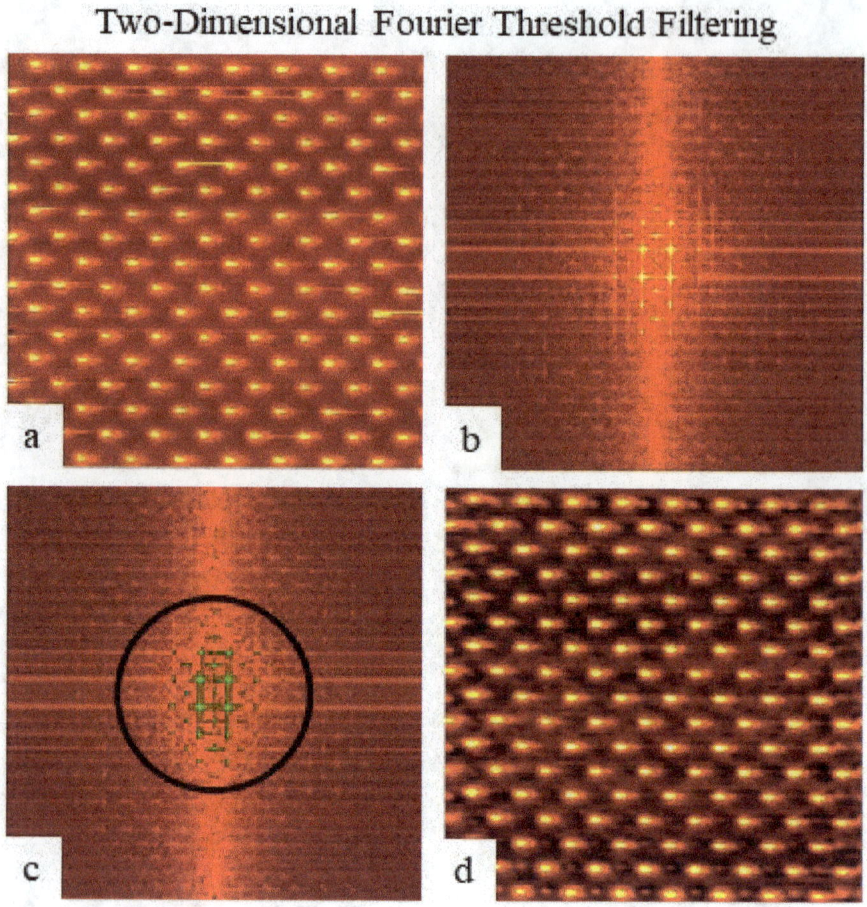

Fig. 2.16 An example of Fourier threshold filtering of the streaky image in (a). The two-dimensional Fourier transform is shown in (b). A threshold was chosen such that only the central peaks would be inverse-transformed. They are the green peaks within the circle in (c). The inverse transform is displayed in (d), where no streaks are evident.

filtering. This is useful, e.g., when it is physically impossible to remove electrical line noise at multiples of 50 or 60 Hz by any other method. Also, some images have strong unwanted components, such as the start-up artifacts (Sect. 2.2.1.8) on the left-hand side of Fig. 2.17a. The frequencies corresponding to the start-up artifacts are stronger than the frequencies forming the features of the sample, a triangular grating, and therefore threshold filtering alters both the artifacts and the desired data.

★ In **Computer Lab 3**, you practice Fourier filtering. ★

The pictures on the left-hand side (a, c, e) of Fig. 2.17 are the real-space images and those on the right-hand side (b, d, f) are the k-space or frequency images of the

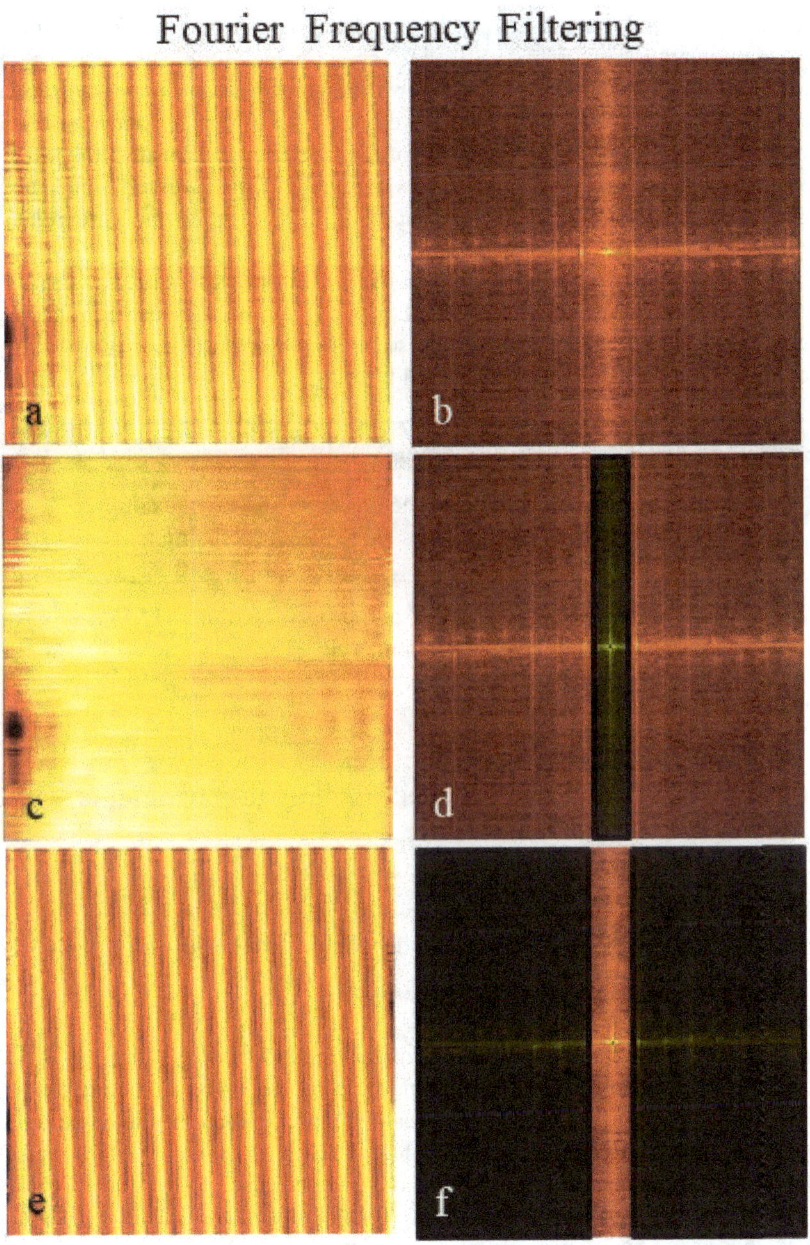

Fig. 2.17 Examples of Fourier frequency filtering of an image of a triangular grating with start-up artifacts on its left-hand edge. The pictures on the left-hand side (a, c, e) are the real-space images. Those on the right-hand side (b, d, f) are the k-space images. The top row (a, b) are the original data, the middle row (c, d) highlight the start-up artifact and slight slope to the original data, and the bottom row (e, f) displays the final image and the frequencies that constitute that image.

triangular grating. Figure 2.17a are the raw data; Fig. 2.17b is the Fourier transform. As a demonstration, in Fig. 2.17d, the k-space values within the green rectangle that compose the start-up artifacts (low frequency values in the x-direction, all frequency values in the y-direction) have been inverse transformed into Fig. 2.17c. These frequencies should be excluded for the final image. If instead the other set of frequencies, those in the green rectangles in Fig. 2.17f, are inverse-transformed, then the resulting image is Fig. 2.17e. Almost all of the start-up artifacts are gone.

To return to the statements in the first paragraph of this section, if there are noise spikes in your image, perhaps from a 50 or 60 Hz source, they can be removed with Fourier frequency filtering. If white noise interferes with the quality of your image, it can be removed with fourier threshold filtering.

AFM Lesson 11 at `youtube.com\AtomicForceMicro`

AFM Lesson 11 sums up the concepts and gives examples of Fourier transforms for image processing.

By now we hope that you have acquired and processed some AFM images yourself. In the next chapter, we present instruments related to AFM, as well as some other common AFM imaging modes.

2.6 Learning objectives, further reading, and problems

! Learning objectives for Chapter 2

From the material in Chapter 2, the reader should now be able to:

2.1 Sketch a diagram of an AFM system.
2.2 Define the error signal.
2.3 Understand the purpose of the feedback circuit.
2.4 Recognize when the feedback gains are too high, about right, and too low.
2.5 Demonstrate image optimization using feedback and other scan parameters.
2.6 Be familiar with common artifacts and how to test and correct for them.
2.7 Describe the common features of a noise spectrum.
2.8 Know how to situate and protect your SPM.
2.9 Understand how Fourier filtering can improve noisy images.
2.10 Acquire and process topographic AFM images.

Further Reading

✓**Tutorial.** Jeon, D., & Willis, R. F. (1991). Feedback system response in a scanning tunneling microscope. Review of scientific instruments, 62(6), 1650–1651.

✓**Book.** Bovik, A. C. (Ed.). (2009). The essential guide to image processing. Academic Press.

✓**Tutorial.** Burnham, N. A., Lyu, L., & Poulikakos, L. (2023). Towards artefact-free AFM image presentation and interpretation. Journal of Microscopy, 291(2), 163–176.

✓**Tutorial.** Delsuc, M. A., & O'connor, P. (2024). The Fourier transform in analytical science. Nature Reviews Methods Primers, 4(1), 49. See also the PrimeView graphical summary based on the article.

? Problems for Chapter 2

2.1 A common set-point force is 60 nN. The type of cantilever that you use for the first time might have a spring constant of 0.18 N/m.
(a) What is the deflection of the cantilever in nanometers?
(b) If the sensitivity of the photodiode is 300 nm/V, what is the set-point voltage?

2.2 The same cantilever as in P.2.1 has a natural frequency of 12 kHz. What is its effective mass? Express your answer in nanograms.

2.3 Figure 2.3 is divided into four regions by gray horizontal lines. Which area(s) (1, 2, 3, or 4) correspond to too high gain, about right gain, and too low gain?

2.4 A typical starting scan range might be 25 μm, with a scan frequency (forwards and backwards) of 0.50 Hz and 128 pixels per line.
(a) What is the distance between pixels? Express your answer in nanometers.
(b) What is the speed of the tip? Express your answer in microns per second.
(c) What is the time per pixel? Express your answer in milliseconds.

2.5 One way to make a quick analysis of noise is to take an error image with zero scan range. Then the "distance" between pixels is actually time. The inverse of time is frequency. If a scan frequency of 5.0 Hz is used on a 128 pixel line, what is the frequency of pixel acquisition? Express your answer in kilo-Hertz.

2.6 Look online for an AFM image that suffers from an artifact. Send the link to your instructor. Be ready to describe it in class in a few sentences.

2.7 A sawtooth wave with no offsets has an amplitude Y and period T. What is its (a) mean value; (b) mean-square value, and (c) root-mean-square value? Hint: Draw sketches of one period of a sawtooth wave (linear up from origin to $T/2$, then a vertical drop from Y to $-Y$, then linear back to t-axis) and its square.

2.8 Find the Fourier coefficients a_0, a_1, a_2, b_1, and b_2 of $\cos \omega t$ using Eqs. 2.5. It is convenient to use a time interval $T = 2\pi/\omega$ equal to one period. You may use a plotting program to demonstrate which coefficients are zero. For any non-zero coefficients, show important mathematical steps leading to numerical answers.

2.9 Lab question. Assuming that you have been able to use an AFM and tried imaging steps by scanning both across and along the steps using different image filters (e.g. raw data, line fit), was there any situation in which the image contrast changed radically? If so, why?

2.10 Lab question. Assuming that you have been able to use an AFM and tried various PI or PID settings, which gains had the most influence over the quality of your images?

2.11 Lab question. Assuming that you have already visited the AFM lab and seen how the AFM is situated and protected, state five noise-reducing aspects of its configuration. If you think its environment could be further improved to reduce noise, also note your suggestions.

2.12 *Challenge Problem.* Derive Eq. 2.7 from Eqs. 2.4 and 2.5.

Chapter 3
Other SPMs and AFM imaging modes

Abstract This chapter develops insights into scanning probe microscopes, specifically scanning tunneling microscopy, lateral force microscopy, and common dynamic modes of atomic force microscopy: the non-contact, intermittent-contact, and off-resonance image acquisition modes. The intent is not to be exhaustive; to do so would require a book in itself (which would quickly become outdated!). Rather, the general ideas are presented and an attempt made to guide the reader's physical intuition.

3.1 Scanning probe microscopy

Atomic force microscopes have many "family members." The family of instruments is known as *scanning probe microscopes* (SPMs), or occasionally *proximal probe microscopes* (PPMs). The first one to be invented was the *scanning tunneling microscope* (STM) in 1981. AFMs followed in 1986. A slew of other variations has since been invented. In the next two sections, STM and a variation that measures friction are discussed. The SPM that ideally measures friction is called the *friction force microscope* (FFM). Some researchers prefer to call them *lateral force microscopes* (LFM), as friction is just one type of lateral force (Sect. 3.3).

This instrument family was introduced in Sect. 1.2.2. To support what was written there, here we add a modification to Figs. 1.3 and 2.1. It is Fig. 3.1, in which we visually emphasize that if a tip-sample interaction can be detected, an SPM can be created...is it your turn to invent one?

> **AFM Lesson 12 at youtube.com\AtomicForceMicro**
>
> AFM Lesson 12 provides an overview of scanning probe microscopy.

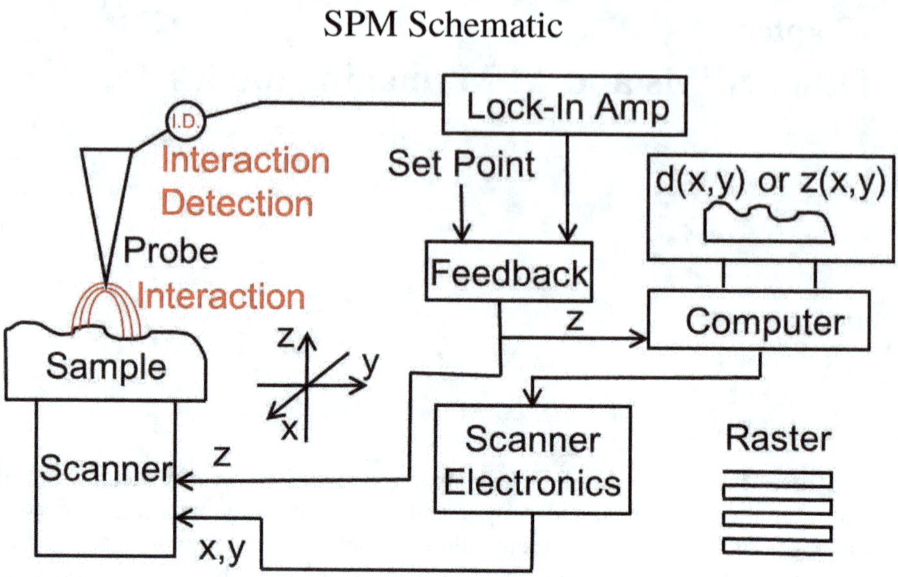

Fig. 3.1 Schematic of a generic Scanning Probe Microscope (SPM). If an interaction between a sharp probe tip and a sample can be detected (red highlights), an SPM can be constructed. All SPMs share the other componenents, as comparison with Figs. 1.3 and 2.1 will reveal.

3.2 Physics of scanning tunneling microscopy

The scanning tunneling microscope (STM) was the first of the family of scanning probe microscopes to be invented (1981); it is historically important. STM is also technically important, as it has mostly been applied to study the surfaces of semiconductors, which are the basis of the electronics and computer industries. With an atomically sharp tip, it allows us to see the atomic structure of materials, the first instrument to do so. To achieve atomic resolution, it uses the phenomenon of "tunneling current," an aspect of quantum mechanics. While quantum mechanics is beyond the scope of this book, an analogy will be provided, so as to provide you with some understanding. Before describing tunneling, however, the required experimental conditions will first be related.

3.2.1 Required experimental conditions

To create an STM image, an STM tip is rastered over a sample surface, just as for an AFM image. However, instead of the deflection of the cantilever being monitored, the "tunneling current" is monitored in order to create an image. Like the "swash" of an ocean wave—a small fraction of the wave—running up a beach, the tunneling current is a small fraction of the current that would be expected if the tip were in

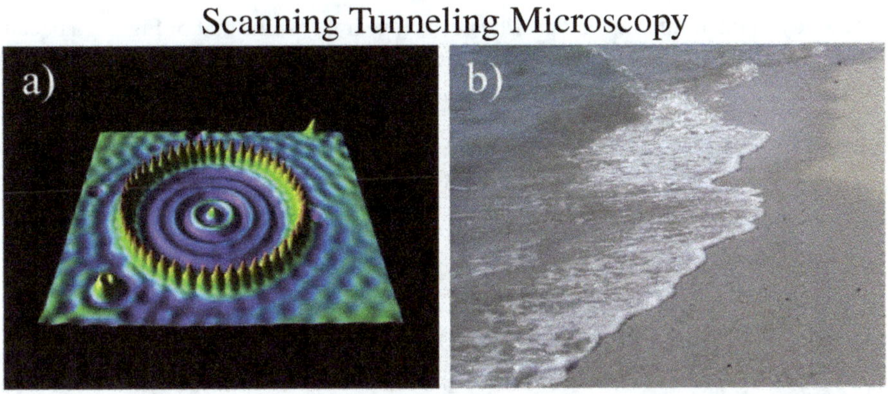

Fig. 3.2 (a) A three-dimensional STM image of a "quantum corral", where the raised structures are individual atoms, and the smaller undulations are electron standing waves. The radius of the corral is about 7 nm. (From Don Eigler, IBM Almaden Research Center) In (b), a swash of an ocean wave is shown; it is the remaining water after the wave has partially climbed the incline of the beach. By analogy, the swash represents the tunneling current used to make STM images. (Photo by Maki Akiyama)

firm conductive contact with the sample. Figure 3.2a displays an STM image of a "quantum corral," where the bumps (mostly in a ring) are individual atoms and the smaller undulations are standing waves of electrons. Figure 3.2b is a photograph of a swash.

Very special circumstances are needed to create images like that of Fig. 3.2a. Firstly, both the tip and sample must be conductors or semiconductors in order to pass a current when a small voltage is applied between the tip and sample. Secondly, because many surfaces naturally form oxides in air, most STM studies are conducted in ultra-high vacuum so that clean surfaces can be prepared. Thirdly, in order to achieve such high lateral resolution, the tip must terminate in just one atom. And in this case, the STM and sample were cooled to approximately 4K in order to minimize atomic mobility and other perturbations. (At room temperature, many atomic species skitter over a surface.)

The atoms of the corral are iron. They are on top of a copper surface. In order to bring the iron atoms into the corral structure, the STM tip was brought close enough to the atoms such that there was a strong force interaction between them. Then the STM tip was displaced laterally, and the iron atoms followed. To image, the tip was drawn back slightly from the surface, where the rastering of the tip to create the image did not displace the iron atoms.

The other aspect of note in Fig. 3.2a is the small undulations on the copper surface. These are electron standing waves, which inside the corral are similar to waves on the surface of an otherwise still pool after you have dropped a pebble into it. Whereas you probably have been introduced to electrons as particles, one fascinating aspect of nanoscale reality is that small particles can also act as waves (although not simultaneously). This is called "wave-particle duality" and is an important theme in the study of quantum mechanics. Physicists love images like that of Fig. 3.2a because the way that the waves behave follows the mathematical formulations of quantum mechanics.

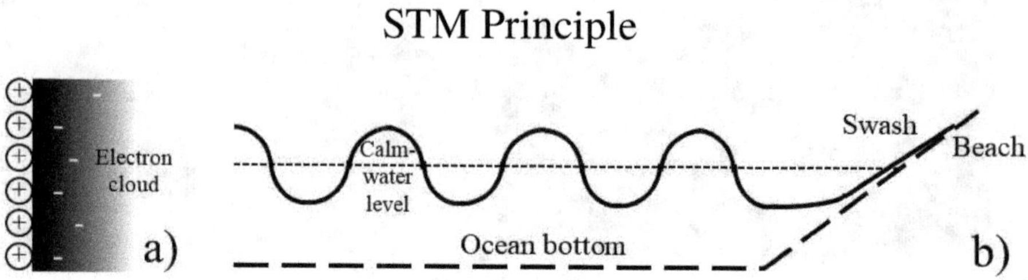

Fig. 3.3 (a) A schematical cross-section of a surface, with the positive nuclei on the left and an electron cloud on the right, where the probability of finding an electron decays rapidly with the distance from the surface. (b) Just as the swash of ocean waves "bleeds" up a beach, the electron cloud bleeds out into space near both the tip and the sample.

3.2.2 What is tunneling?

In this section, we attempt to describe tunneling without first providing a course on quantum mechanics. (Although you are certainly encouraged to study it at some point!)

Tunneling stems from the nanoscale properties of surfaces and atoms. Atoms are composed of positively charged centers (nuclei) and negatively charged electrons. Electrons have much less mass than the nuclei. Consequently, they spread out in space, a result of a quantum-mechanical effect called the Heisenberg Uncertainty Principle.

As applied to surfaces, this means that there is a layer of negative electron charge covering the outermost positive nuclei, similar to two sheets on a bed. Unlike a sheet, however, the density of the electron charge is high close to the bed and low far away from it. Rather, it is more like a cloud or a swash on a beach. A cross-sectional schematic of a surface is shown in Fig. 3.3a. The electric double layer formed by the positive and negative layers means that it is difficult for electrons inside the sample to leave the sample, as if they need to climb up an energy barrier. It requires a few electron volts of energy to extract them. This energy is called the *work function*.

The tunneling phenomenon is similar to ocean waves and their swash, depicted Fig. 3.3b. The swash is what remains as the waves climb up the beach. Similarly, the tunneling current in an STM is what leaks out of the energy barrier of the surface. The term "tunneling" refers to how the current "tunnels" through the energy barrier of the electric double layer.

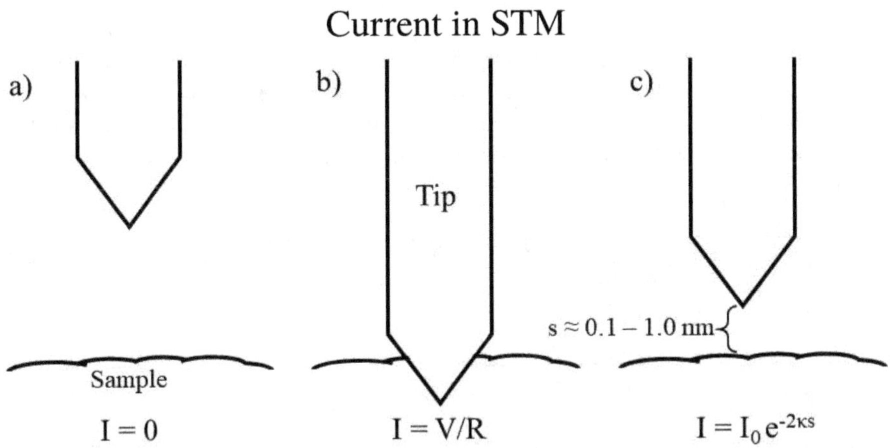

Fig. 3.4 The current I expected for three different tip-sample separations or indentations. (a) Greater than a few nanometers separation, no current flows between the tip and sample. (b) When the tip is firmly embedded in the sample, the current follows an Ohm's Law (V/R) relationship. (c) When the tip and sample are separated by a distance *s* of about 0.1 to 1 nm, measurable tunneling current can be detected.

3.2.3 Distance dependence of tunneling

When a tip is far from a sample surface with a small voltage difference applied between them, there is no contact, and no current can flow, as sketched in Fig. 3.4a. If the tip is embedded in the sample, then current will follow Ohm's Law, $V = IR$, rewritten in Fig. 3.4b as $I = V/R$. In Fig. 3.4c, the situation for a tunneling current is shown.

Let us now describe the tunneling effect mathematically. The electron cloud in Fig. 3.3a decays exponentially with distance from the sample surface. Similarly, an electron cloud is associated with the STM tip, and it too decays exponentially with distance. We call the tip-sample separation s, and the tunneling current I has the form

$$I = I_0 e^{-2\kappa s}, \tag{3.1}$$

where I_0 is the current when $s = 0$. The constant κ indicates how quickly the respective electron clouds decay with distance. Because the argument of an exponent (or trig function, for that matter) must be dimensionless, we immediately see that κ is measured in inverse distance. Such quantities are called wavenumbers. Its mathematical definition is

$$\kappa = \frac{1}{\hbar}\sqrt{m_e(\Phi_t + \Phi_s)}, \tag{3.2}$$

where $\hbar$ is Planck's constant divided by 2π, m_e is the mass of the electron, and Φ_t and Φ_s are the work functions of the tip and sample, respectively. Some of the problems at the end of this chapter ask you to calculate various values of κ and I so that you appreciate the rule of thumb for tunneling: every Angstrom of separation leads to a factor of ten decrease in current. This high sensitivity to distance is the reason for STM's incredible lateral resolution: when a tip is terminated with just one atom, almost all of the current passes through it, and the tip can follow the features of the sample atom by atom.

AFM Lesson 13 at youtube.com\AtomicForceMicro

AFM Lesson 13 gives a video introduction to STM.

3.3 Lateral force microscopy

After motivating the reader for the material in this section, a short review of friction physics is provided. It is followed by descriptions of the types of lateral force data and friction loops. The last subsection demonstrates how analysis of friction loops is conducted, within several limiting assumptions.

3.3.1 Motivation for lateral force microscopy

Would you buy a car that would neither stop nor go? Maybe not! Friction between the brake pads and the wheels stops a car. Friction between the tires and the road pushes the car forward. At the same time, we wish we could lower the friction among the moving parts of the engine and drive train, so as to improve gas mileage.

If an AFM has a quadrant photodectector, friction and other lateral forces can be detected by the horizontal movement of the reflected light spot on the photodetector due to twisting of the cantilever; please review Fig. 1.1. This mode of operation in AFM is known as lateral force microscopy (LFM), or to emphasize the interest in friction itself, friction force microscopy (FFM). Because friction can also influence the vertical movement of the reflected light spot when the tip is in contact with the sample, data interpretation is simplified if the fast-scan direction is perpendicular to the long axis of the cantilever. Then horizontal motion of the reflected light spot at the photodetector is predominantly due to lateral forces and vertical motion to normal forces.

The advantage of LFM studies of friction is that one can model the tip-sample interaction as a single point of contact between the tip and sample. In contrast, in studies of macroscale objects such as tires and roads, which are inherently rough (at least from the perspective of nanoscience!), the surfaces will contact each other at many different contact points, each with a different set of characteristics.

3.3.2 Review of friction physics

You probably learned some equations for friction in first-year physics. The classic one is $F_k = \mu_k N$, where F_k is the kinetic or sliding friction, μ_k is the kinetic friction coefficient, and N is the normal force, "normal" meaning perpendicular to the surface. The friction coefficient μ is a unitless positive coefficient, usually in the range of 0 to 1. While the equation for F_k does not hold in every situation, it is a simple one that applies in many circumstances, and we thereby limit the discussion here to this simple case.

Static friction is described somewhat differently. You might have noticed that getting objects to start sliding often requires more force than to keep an object sliding once it is moving; this is attributed to the details of how the two interlocking surfaces interact. So we expect that the static friction coefficient, μ_s, to be higher than μ_k, and indeed in general this is true. However, there is more inconsistency in static friction measurements than in sliding friction measurements. Yet repeated static friction measurements will usually yield an upper limit for static friction and hence μ_s. Thus the equation for static friction is expressed as $F_s \leq \mu_s N$. Note that instead of an equals sign, there is a "≤," which is the reminder that a static friction measurement is not as repeatable as a kinetic friction measurement.

3.3.3 Lateral force data

Lateral force data come in two forms. The first is LFM images, where at the same time the feedback circuit is holding the cantilever deflection nominally constant and the topographic data continue to be acquired, simultaneously the twisting of the cantilever is acquired and plotted. An example of simultaneously acquired AFM and LFM images is given in Fig. 3.5, where the top images (a, b) are AFM's topographic forward and reverse data, and the bottom ones (c, d) are LFM's forward and reverse lateral force data of an incompletely exfoliated pressed compact disk. The lighter-colored topographic level corresponds to aluminum that did not separate from the polycarbonate substrate. For the lateral force images, the extremes in colors (that is, very light or very dark) represent higher lateral force magnitudes for the two twisting directions.

The second type of data is what are known as "friction loops." For a given fast-scan line, the LFM signal is acquired and plotted in the same way as a fast-scan topographic cross-section. Whereas for a topographic cross-section hysteresis between the back-and-forth motion of the tip is ideally minimal, for a friction loop, significant differences are expected because the tip twists in one direction on the forward scan and in the other on the backward scan. As you will see below, the features of the friction loop can be related to the slope of the sample and the friction.

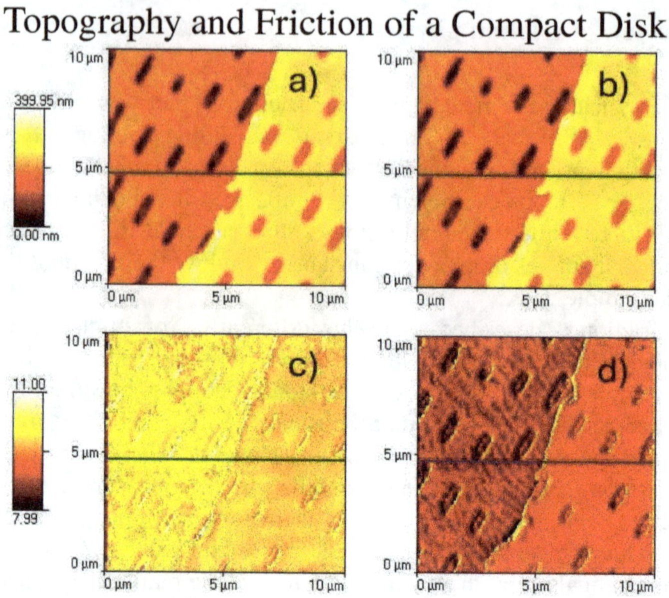

Fig. 3.5 Simultaneously acquired topographic (a, b) and lateral force (c, d; uncalibrated color scale) images of an exfoliated pressed compact disk, where the depressions are the bits. The left-hand images (a, c) were acquired as the tip was scanning in the forward direction (left to right) and the right-hand images (b, d) as the tip was scanning in the backward direction (right to left). The higher level on the right of the topographic images is aluminum that remained after the top surface of the compact disk was removed to reveal the aluminum and polycarbonate substrate. The horizontal gray lines just below 5 µm on the y-axis show where you will be asked to sketch a topographic cross-section and a friction loop in P. 3.8.

3.3.4 Friction loops

Why do we keep talking about lateral forces instead of friction forces? Because the tip can be twisted by not only friction at its very apex, but also by steps on the sample that contact the tip on its side, instead of on its apex. Also, on sloped surfaces, a component of the normal force points horizontally instead of vertically, as will be detailed below.

To become familiar with friction loops, please first look at Fig. 3.6a. It is a schematic cross-section of a surface with two steps, a slope, and two different friction coefficients, as if the surface is composed of two different materials. For visual clarity, we are dropping the subscripts that denote kinetic or sliding friction (Sect. 3.3.2) on the friction coefficients μ. Imagine the tip scanning back and forth across the cross-section.

Now examine Fig. 3.6b. It is drawn directly below Fig. 3.6a to the same horizontal x-scale. As it encounters steps and slopes, the amount of twisting of the cantilever, V_{LFM}, is recorded at the same time the vertical deflection of the cantilever is held constant. As the tip can twist both to the left and to the right, V_{LFM} can be positive or negative. The height difference F_{diff} between the back-and-forth motion during a scan is related to the amount of kinetic (sliding) friction. Note that there are two different "thicknesses" of the loop, corresponding to the two different friction coefficients. The average value of a section of the loop, F_{av}, is related to the angle of the slope; the steeper the slope, the greater the magnitude of F_{av}. It can be positive or negative. Spikes occur at steps and sometimes at the beginning of a scan.

The vertical axis in Fig. 3.6b is a voltage instead of a force. Why? It is challenging to calibrate the torsional spring constant of a cantilever. It is high compared to the spring constant for vertical deflection (the "normal" spring constant) because the cross-sections of cantilevers are much wider than they are thick. Equations for spring constants and approaches for calibrating normal spring constants are described in Sect. 4.1. Researchers have come up with a vast array of methods to calibrate torsional spring constants; these are beyond the scope of this book.

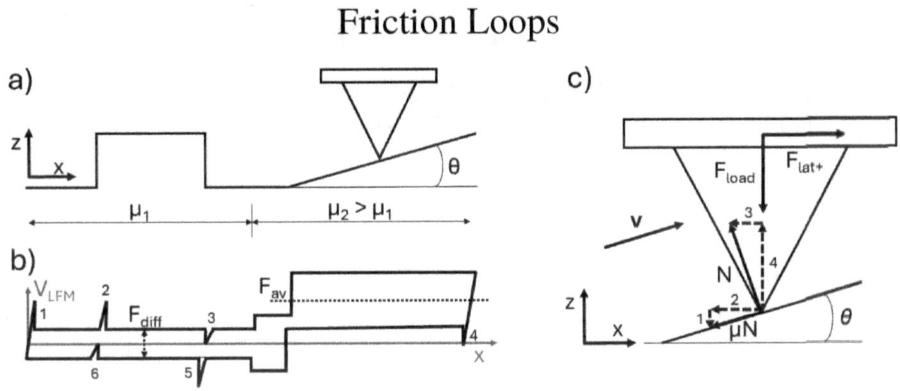

Fig. 3.6 Schematics for understanding friction loops. (a) Cross-section of a sample with two different kinetic friction coefficients μ; the k subscript on μ has been dropped for clarity. The view is down the long axis of the cantilever. The coordinate directions are shown. (b) The corresponding friction loop: the voltage corresponding to lateral forces on the tip, V_{LFM}, as a function of x-position along the sample. The physical significance of the distance F_{diff} and the y-value F_{av} will be derived in the text. In P. 3.9, the reader will be asked about the spikes numbered 1–6. (c) The forces acting on the tip from the cantilever and sample as the tip travels at constant velocity **v** up a slope at angle θ above the horizontal. In P. 3.11, the reader will be asked about the force components labeled 1–4.

3.3.5 Lateral force analysis

To learn what F_{diff} and F_{av} represent, we start with a sketch of the forces acting on the tip from the cantilever and sample, Fig. 3.6c, as the tip is moving up a slope at angle θ above the horizontal at constant velocity **v**. F_{load} is the vertical deflection of the cantilever times its normal spring constant. F_{lat+} is the force from the cantilever that moves the tip laterally up the slope due to the scanner exerting a horizontal force on the cantilever. N is the normal (i.e., perpendicular) force from the surface, and μN is the friction. The coordinate directions are shown, with x perpendicular to the long axis of the cantilever and z pointing vertically.

The assumptions in the following derivation are: the tip moves at constant velocity, the twisting of the tip is negligible but measurable, the cantilever is horizontal, the sample is horizontal in the y-direction, adhesion is negligible, and friction is a linear function of the normal force. So many! But still, the following equations lead to insights about friction loops.

The assumption about constant velocity means that the tip is in equilibrium, in other words, that the forces on it sum to zero. By breaking the forces into their components (dashed lines in Fig. 3.6c), and accounting for the directions in which they point, the following equilibrium equations can be written.

$$\sum F_x = 0 = F_{lat+} - N\sin\theta - \mu N\cos\theta, \quad (3.3a)$$

$$\sum F_z = 0 = -F_{load} + N\cos\theta - \mu N\sin\theta. \quad (3.3b)$$

In P. 3.11, you will be asked to write the companion equations for a tip traveling down the slope. And in P. 3.17, you will be asked to solve for F_{lat+} and F_{lat-}. The solutions are these:

$$F_{lat+} = F_{load}\left(\frac{\tan\theta + \mu}{1 - \mu\tan\theta}\right), \quad (3.4a)$$

$$F_{lat-} = F_{load}\left(\frac{\tan\theta - \mu}{1 + \mu\tan\theta}\right). \quad (3.4b)$$

Friction coefficients tend to be on the order of 0.2–0.3 for many tip-sample combinations in air. Sample slopes in AFM experiments are also usually less than 45°, such that their tangents are less than one. Thus the $\mu\tan\theta$ term in the denominator is sometimes much less than one, can be considered negligible as compared to one, and thus can be ignored. This additional assumption simplifies the above equations to

$$F_{lat\pm} = F_{load}(\tan\theta \pm \mu), \quad (3.5)$$

where the "+" in the "±" sign represents the solution for the tip traveling up the slope, and the "−" sign for down the slope. By taking the average and difference of the two situations, the useful results for interpreting friction loops are

$$F_{av} = F_{load}\tan\theta, \quad (3.6a)$$

$$F_{diff} = 2\mu F_{load}. \quad (3.6b)$$

Have another look at Fig. 3.6b and imagine how it will change depending on the load, angle, and friction coefficient.

AFM Lesson 14 at youtube.com\AtomicForceMicro

AFM Lesson 14 describes lateral force microscopy.

3.4 Other AFM modes of operation

Thus far, we have discussed only a few modes of operation in AFM: contact mode, lateral force microscopy, and friction loops. If you are following the laboratory exercises associated with this book, you have been using contact mode. Now it is time to expand your experience to other operational modes. Two common imaging modes, the non-contact and intermittent-contact modes, require a basic understanding of resonance. In both modes, the cantilever is oscillated near resonance. Resonance is thus introduced forthwith, yet a detailed mathematical presentation of resonance is deferred until Chap. 7. Another mode of operation occurs at low frequencies, and it is also described after the resonance approaches. Collectively, these are referred to as dynamic modes.

3.4.1 Introduction to resonance

An object, when struck, will vibrate at its natural frequency. Think of a bell. Or when a swing is released from a high position, it will oscillate. When you drive a swing by shifting your weight back and forth, you incur resonance of the swing. You can only reach high amplitudes on the swing if your shifts in weight correspond to a frequency that closely matches its natural frequency.

A plot of the amplitude and phase angle as a function of frequency of a generic oscillator—such as an AFM cantilever—is shown in Fig. 3.7a. Its natural frequency is 10, and thus its amplitude is highest near 10. Close to zero frequency, this oscillator responds with the same amplitude as its input; thus its amplitude is here denoted as 1. At high frequencies above resonance, its amplitude is below 1.

What is meant by phase angle? The phase angle shows how the oscillator's response is related to cyclical excitation in time. A phase angle of zero means that the response and excitation are synchronized. A phase angle of $-180°$ means that the response is half a cycle behind the excitation. The phase angle passes through $-90°$ at the oscillator's natural frequency. The minus sign shows that the response lags the excitation.

To understand how resonance enables AFM imaging, two equations must be introduced below; you have likely seen them if you have taken a course on differential equations. The first tells you the amplitude of the free end of the cantilever, c_1, as a function of the oscillatory excitation force F_1, the cantilever's spring constant k_c, the radial driving frequency ω, the cantilever's effective mass m, and the damping of the cantilever, b_c. We have not mentioned damping before. A common source of damping is that of air or water, such as you feel on your arm when you move your hand quickly back and forth. Any energy loss will increase the damping. The second equation is for the tangent of the phase angle θ. You see that it is dependent on the same set of parameters, except for the excitation force.

$$c_1 = \frac{F_1}{\sqrt{\left(k_c - \omega^2 m\right)^2 + (\omega b_c)^2}},$$
$$\tan\theta = \frac{-\omega b_c}{\left(k_c - \omega^2 m\right)}. \tag{3.7}$$

With these two equations, we can understand how forces from a sample influence the behavior of an AFM probe.

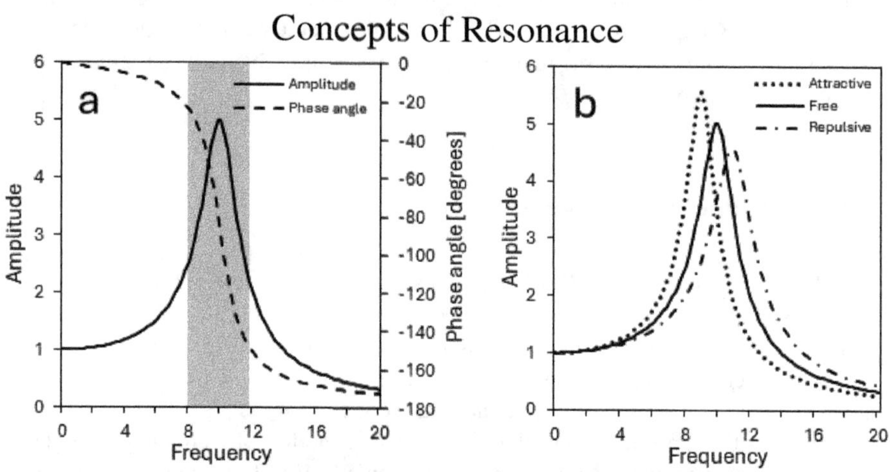

Fig. 3.7 Concepts of resonance. (a) Schematic of amplitude and phase angle (dashed line) as a function of frequency for an arbitrary cantilever or other oscillator, such as a spring or swing. At low frequencies, the amplitude of oscillation of the response matches that of the excitation, denoted here as 1. Near resonance (shaded area), the same excitation amplitude leads to larger response amplitudes as the phase angle changes from being in phase (0° or 0 radians, excitation and response moving together) to being out of phase (−180° or $-\pi$ radians, excitation and response moving in opposite directions). The minus sign indicates that the response lags the excitation. At high frequencies, response amplitudes are small. (b) If a cantilever free of influence (same as in a), solid line) is affected by an attractive (dotted line) or repulsive force (dot-dashed line) and its damping remains the same, the peak position shifts in both maximum amplitude and frequency. The phase responses maintain their shape but shift in frequency; they are not shown here for ease of inspecting the changes in amplitude.

3.4.2 Why resonance changes

In the presence of an attractive or repulsive force, an oscillator's resonance properties change. Figure 3.7b shows how. Under the influence of an attractive force, its amplitude increases and the position of its maximum shifts to lower frequency. Conversely, under the influence of a repulsive force, an oscillator's amplitude decreases and it shifts to the right. Why?

Imagining now the tip of an AFM probe oscillating near an attractive surface (e.g. electric charges of different sign on the tip and sample), the tip feels a downward force from the surface in the opposite direction to the restoring force of the cantilever. The net force on the tip is reduced as compared to the situation with no surface force, as if the spring constant of the cantilever were reduced in the relation $\mathbf{F} = -k_c\mathbf{d}$. Thinking mathematically, a smaller spring constant causes the first squared term in the denominator of the c_1 equation above to equal zero at a lower value of ω. And the second squared term in the denominator is smaller at that value of ω than it would have otherwise been, increasing the value of c_1. Similarly, a plot of $\tan\theta$ shifts to lower frequency, although it retains its shape.

Conversely, when the tip of an AFM probe is affected by a repulsive force from a surface (e.g. electric charges of the same sign on both), the net force pushing and pulling the tip upwards is greater in magnitude than just the restoring force of the cantilever. It is as if the cantilever has a larger spring constant. The two terms in the denominator of the c_1 equation evolve in the opposite sense, with an increase in frequency of the maximum amplitude and a decrease in its value. No surprise—a plot of $\tan\theta$ shifts to higher frequency under the influence of a repulsive force from the sample.

Another way of understanding the influence of forces on the behavior of an AFM cantilever is with the use of potential diagrams. We model the cantilever as a spring, with potential $U = ½k_c d^2$. Figure 3.8 shows three spring potentials as a function of distance. The free, unperturbed cantilever's potential is the solid line; the attractive situation with a lower spring constant is the dotted line, and the dot-dash line represents the potential for the tip subject to a repulsive force (higher spring constant). The total energy E_{tot} of the cantilever is the sum of its kinetic and potential energies, $E_{tot} = K_E + U$; it is represented by the solid horizontal line. (Remember that $K_E = ½m\vartheta^2$, with m the effective mass and ϑ its velocity.) You see that it intersects the potential diagrams at different positions—the cantilever's turning points. A turning point on a potential plot corresponds to the cantilever's amplitude of oscillation.

For all three cases, the kinetic energy is the difference between the total energy and the potential energy, $K_E = E_{tot} - U$. For all three cases, it is at a maximum at position 0, represented by the vertical solid line labeled K_{max}. Since the maximum kinetic energy is the same for all three cases, yet the distance of travel (twice the amplitude) is different, the oscillation frequencies change accordingly. For the attractive case, it takes longer for the cantilever to move farther, and hence the frequency goes down. For the repulsive case, the smaller amplitude means the cantilever's frequency goes up.

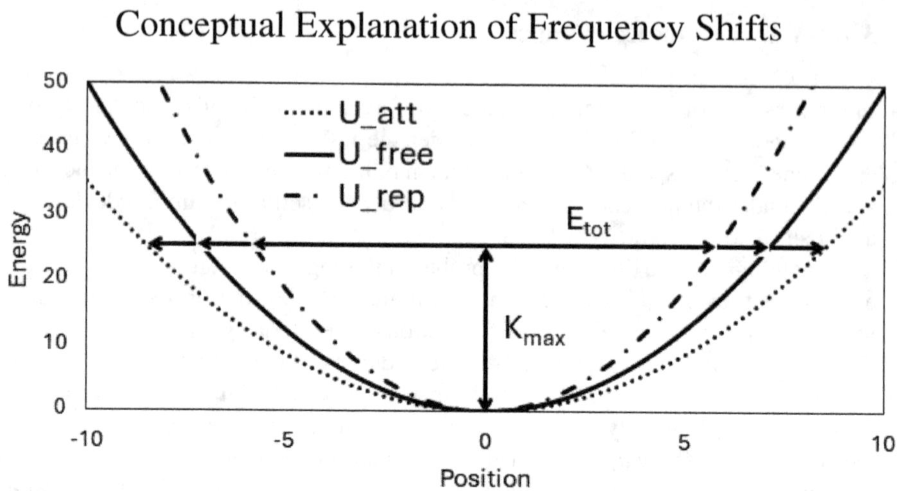

Fig. 3.8 Conceptual explanation of frequency shifts. An oscillator's potential energy (U) diagram follows $U = \frac{1}{2} k_c d^2$, with k the spring constant and d the position, or in the case of a cantilever, the deflection. Attractive forces make the effective spring constant smaller and repulsive forces make the effective spring constant larger. As a result, the attractive potential, U_{att}, rises more gently than the potential for the free oscillator, U_{free}, which in turn rises more gently than the potential for the repulsive potential, U_{rep}. The horizontal line labeled E_{tot} indicates the total amount of energy of the oscillator, such as that supplied to the base of an AFM cantilever. The total energy is the sum of the kinetic and potential energies, $E_{tot} = K_E + U$. At position 0, the potential is 0 and the oscillator's maximum kinetic energy (K_{max}, vertical line) is maximum. The maximum kinetic energy is the same for all three cases shown here. The turning points of the oscillator are shown by the arrows on the total energy line E. Because the kinetic energy is the same at position 0 for all three cases, yet the turning points and thus amplitudes of oscillation differ, the frequency of the oscillator must change in order to keep the maximum kinetic energy constant.

● In **MacroLab 3**, you see what causes the amplitude and frequency of an oscillating cantilever to shift. ●

3.4.3 Near resonance: non-contact and intermittent-contact modes

Now that you have an idea of resonance behavior and how resonance changes upon the influence of an external force from the two sections above, let us introduce two dynamic modes: non-contact mode and intermittent-contact ("tapping") mode. Both use resonance changes to control the height of the scanner during imaging. As already mentioned in Sect. 2.1, instead of using the deflection of the cantilever as the feedback circuit's set-point parameter, a lock-in amplifier determines the cantilever's oscillation amplitude and sends it on to the rest of the feedback circuit.

Figure 3.9 demonstrates the idea. A typical tip-sample interaction is attractive upon approach of the tip towards the sample, then becomes net repulsive as the oscillating tip contacts the sample every cycle as the tip-sample distance is further decreased. The resonance peak thus first shifts to lower frequencies, then to higher ones, denoted by the curved arrow at the top of the figure. For non-contact AFM (NC-AFM), the drive frequency of the cantilever is chosen on the upper side of the resonance peak, such that the set-point amplitude is reached (filled circle) while the tip and sample are still relatively far away. For intermittent-contact AFM (IC-AFM), the drive frequency of the cantilever is on the lower side of the resonance peak, such that when the resonance peak has shifted to a net repulsive position greater than free resonance, the set-point amplitude (filled diamond) can be met. Typical set-point amplitudes for IC-AFM are 60–80% of the free amplitude.

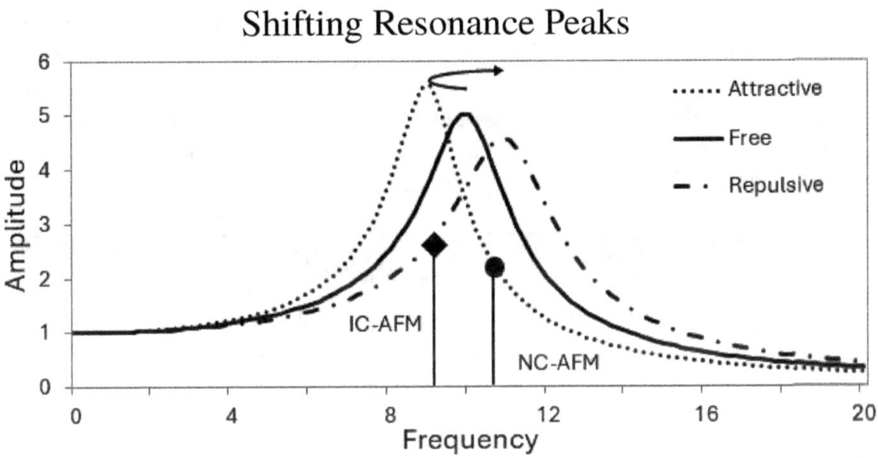

Fig. 3.9 Representative schematic resonance peaks for attractive (dotted line), free (solid line), and repulsive (dot-dash line) tip-sample interactions. The curved arrow above the peaks indicates a typical sequence during tip-sample approach: first the resonance shifts to lower frequencies under the influence of attractive forces from the sample, then shifts to higher frequencies if the tip-sample separation is lowered enough that intermittent contact is achieved. Set-point amplitudes are chosen to be less than the free-cantilever resonance amplitude. Therefore, if intermittent contact is desired, a frequency below that of free resonance is chosen (left vertical line with diamond), and if non-contact is desired, a frequency above that of resonance is chosen (right vertical line with circle).

An assumption in Figs. 3.7–3.9 is that the cantilever potential remains lightly perturbed. The reader is cautioned that this is not always true! Chapter 7 provides some insights as to what happens when perturbations become strong. For this chapter, we limit the discussion and mention that more than one amplitude of oscillation is possible for a given tip-sample situation, possibly leading to instabilities (sudden changes) in apparent height while imaging. These do not occur for IC mode at set-point amplitudes less than 100%, and thus IC-AFM is more commonly used than NC-AFM. However, if tip contamination and tip stability are concerns, NC-AFM is appropriate.

While in Fig. 3.9 the non-contact drive frequency is shown as being on the upper side of the free resonance peak, it should be mentioned that non-contact mode can also be achieved at the free resonance frequency with set-point amplitudes somewhat below 100%. In this case, some manufacturers offer the use of the phase angle as the feedback parameter; the phase angle changes dramatically around resonance.

It's not a bug—it's a feature! Because the phase angle changes so dramatically near resonance, it is often simultaneously monitored during IC-AFM topography maps; this is called phase imaging. Strong phase contrast is often observed from one part of a sample to another. Phase angles become greater in magnitude for more adhesive, viscous, and compliant areas of surfaces.

In addition to instabilities, a challenge of near-resonance data acquisition is that the relationship between force and set point is non-linear. Another challenge is that the resonance frequency is dependent on temperature, the density of the medium in which the cantilever oscillates, and to the extent of the air-damping between the cantilever and sample, which depends on the tip-sample separation. Small changes in the environment during a lab session can shift the resonance peak sufficiently to change the imaging conditions, without the operator becoming aware of the change. To avoid complications due to resonance phenomena, off-resonance acquisition has become popular in recent years.

> **AFM Lesson 15 at `youtube.com\AtomicForceMicro`**
>
> AFM Lesson 15 introduces an AFM's dynamic modes of operation.

3.4.4 Off-resonance data acquisition

To avoid the challenges associated with near-resonance data acquisition, the cantilever can be oscillated at low frequencies (say, a few kHz), far from resonance. Like in NC- and IC-AFM, the cantilever is driven sinusoidally, but here the tip is in contact with the surface for a significant fraction of the sinusoidal cycle. Amplitudes are typically low enough to hold the tip-sample force to below a few nano-Newtons.

Whereas in contact mode the deflection of the cantilever is used as a set point, and in non-contact and intermittent-contact modes the amplitude of oscillation is the set point, in off-resonance data acquisition, the highest force during a sinusoidal cycle is the set-point parameter. The trace of the tip's motion during one cycle is called a force curve. Chapters 5 & 6 delve into force curves and how mechanical properties can be extracted from them. A standout advantage is that in this mode mechanical properties can be mapped while also acquiring topography.

It is a relatively new approach, as it relies on fast detection electronics that did not exist in early versions of AFMs. Different AFM manufacturers use different terms to describe this data acquisition mode: for example, peak-force tapping, off-resonance tapping, wave mode, and fast force mapping.

Further advantages and disadvantages of dynamic AFM imaging modes are presented below.

3.4.5 Advantages and disadvantages of AFM imaging modes

A new AFM user might have trouble deciding which image acquisition mode to first attempt. In this section, we briefly summarize the characteristics of contact, non-contact, intermittent-contact, and off-resonance image acquisition modes and highlight their advantages and disadvantages. The reader should refer to the previous two sections of this chapter if a deeper understanding is needed.

Contact mode. Contact-mode AFM is the easiest to understand and implement. The user can scan to collect an image, by monitoring either the deflection of the cantilever, keeping the vertical position of the scanner at constant height, or by engaging the feedback circuit using the cantilever's deflection as the set-point parameter. In this latter case, the force is nominally the same throughout the image. If the twisting of the cantilever is also monitored, friction maps can be collected simultaneously with topography. The user can stop scanning to generate force curves at specific points to help understand the sample's mechanical properties (see Chaps. 5 & 6).

While attractive due to its relative simplicity, the high pressures generated underneath the very small AFM tips almost always immediately change the geometry and chemistry of the tip and also risk damaging the sample. Furthermore, if the objects of interest are lightly adhered to the sample substrate (such as DNA, carbon nanotubes, or other absorbates), the tip can push them out of the way instead of imaging them.

Non-contact mode. The lack of tip-sample contact (at least, nominally!) is appealing for imaging delicate samples and lightly adhered absorbates and for minimizing contamination or damage to the tip or sample. In NC-AFM, the cantilever is oscillated at or slightly above its resonance frequency and its amplitude is monitored as the set-point parameter.

The need to track oscillations for several cycles in order to determine the amplitude slows down image acquisition. And if the cantilever is not stiff enough relative to the tip-sample interaction, it can jump from one amplitude to another, as will be discussed in Chap. 7. The amplitude in NC-AFM is usually limited in order to minimize this disruption. But by limiting the amplitude, if the sample is adhesive, and the tip does happen to touch it, there might not be enough restoring force of the cantilever to break contact.

Intermittent-contact mode. In this mode, jumps in amplitude are unlikely with set-point amplitudes less than 100% of the free oscillation amplitude. Like non-contact mode, the amplitude of oscillation is the set-point parameter. The drive frequency is set to be slightly below resonance. IC-AFM works well for adhesive and viscous samples, since the typically large amplitudes mean that the restoring force of the cantilever breaks the adhesive contact. It also is a good choice for samples with significant changes in height since the large amplitudes bring the tip above most features for part of the oscillation cycle. It reveals contrast in materials properties if the phase angle is simultaneously acquired.

Like NC-AFM, IC-AFM tends to be slower in image acquisition so that the system has time to determine the cantilever's oscillation amplitude. The tip-sample force and pressure are non-linear functions of the acquisition parameters; it can thus be difficult to estimate if there might be tip or sample damage in advance of imaging. Modeling of the phase and amplitude and their dependence on sample properties has revealed that the mathematical solutions are not unique, and therefore one should be cautious in interpreting phase images, keeping in mind not only non-uniqueness, but also the sensitivity of the resonance frequency and phase to changes in environment.

Off-resonance mode. Advantages of off-resonance acquisition, when the cantilever is oscillated with small amplitudes at frequencies of a few kiloHertz, are: (i) sensitivity to the resonance shifts of the near-resonance modes is avoided, (ii) the tip-sample force is linear with oscillation amplitude and can be kept small, (iii) sample properties can be extracted from the analysis of the real-time trace of the tip's motion and those properties mapped along with the topography, and (iv) like the other dynamic modes, weakly adhered absorbates may be easier to image than in contact mode.

Disadvantages are that (i) data acquisition can be fairly slow with the low excitation frequency, (ii) the typically low oscillation amplitude may not be enough for sticky samples and large changes in feature height, and (iii) the required advanced electronics can be expensive.

Now at the end of the third chapter of this text, I hope you have started to grasp some of the basic principles of AFM. I also hope that you have been able to acquire some of your own data. New AFM users frequently ask about how AFM probes are designed and manufactured and also about how AFMs are calibrated. The next chapter addresses these aspects of this technology.

3.5 Learning objectives, further reading, and problems

! Learning objectives for Chapter 3

From the material in Chapter 3, the reader should now be able to:

3.1 Compare and contrast an SPM, an STM, an AFM, and an LFM.
3.2 Discuss STM's experimental requirements.
3.3 Describe by analogy what tunneling current is.
3.4 Know how static, sliding, and average friction are determined from a friction loop and what influences them.
3.5 Given the topography and friction coefficients of a hypothetical surface, sketch the expected friction loop.
3.6 Distinguish among contact, non-contact, intermittent-contact, off-resonance, and lateral-force modes of AFM operation.
3.7 Know the advantages and disadvantages of the different AFM modes of operation and understand when they are used.

Further Reading

✓**History.** Binnig, G., Rohrer, H., Gerber, C., & Weibel, E. (1982). Surface studies by scanning tunneling microscopy. Physical review letters, 49(1), 57.

✓**History.** Ogletree, D. F., Carpick, R. W., & Salmeron, M. (1996). Calibration of frictional forces in atomic force microscopy. Review of Scientific Instruments, 67(9), 3298–3306.

✓**Tutorial.** Dufrêne, Y. F., Ando, T., Garcia, R., Alsteens, D., Martinez-Martin, D., Engel, A., ... & Müller, D. J. (2017). Imaging modes of atomic force microscopy for application in molecular and cell biology. Nature nanotechnology, 12(4), 295–307.

✓**Tutorial.** Bian, K., Gerber, C., Heinrich, A. J., Müller, D. J., Scheuring, S., & Jiang, Y. (2021). Scanning probe microscopy. Nature Reviews Methods Primers, 1(1), 36.

? Problems for Chapter 3

3.1 In Fig. 3.1, a generic tip-sample interaction is shown. What is the interaction for (a) AFM? (b) STM? (c) LFM? (d) What other kinds of interactions could there be?

3.2 Look online for a variation of an SPM that is not mentioned in Sect. 3.1. Choose one, and send its acronym and the link to your instructor. Be prepared to describe it in class in a few sentences.

3.3 Look online for cool STM images. Choose one, and send the link to your instructor. Be prepared to describe the experiment in class in a few sentences.

3.4 (a) Show that the units on both sides of Eq. 3.2 are the same, using Joules. (b) Now do the same thing using electron volts.

3.5 (a) Calculate the wavenumber κ of an electron if $\Phi_s = 2.00$ eV and $\Phi_t = 5.00$ eV. Use Joules. Express your answer in 1/nm. (b) Do you get the same answer if you use electron volts? (c) If $I_0 = 10$ µA, what is I when $s = 0.10$ and 1.0 nm?

3.6 (a) Calculate the wavenumber κ of an electron if $\Phi_s = 5.00$ eV and $\Phi_t = 5.00$ eV. Express your answer in 1/nm. (b) If $I_0 = 10$ µA, what is I when $s = 0.10$ and 0.20 nm? Express your answers in µA. (c) How many significant figures should be given in the answers to (a) and (b)? (d) Why?

3.7 (a) Figure 3.5 displays AFM and LFM images of polycarbonate and aluminum after exfoliating the top surface of a pressed compact disk. The polycarbonate corresponds to the lower level of the topographic images. Which material has the higher friction coefficient? How can you tell? (b) Are there start-up artifacts? (c) If so, why do they occur where they do?

3.8 Sketch a cross-section of the topography along the gray horizontal line just below 5 μm in Fig. 3.5. Directly below your topo sketch, sketch what the friction loop looks like, based on the lateral-force images in Fig. 3.5.

3.9 (a) In Fig. 3.6b, what are the spikey features labeled 1–6? Why do some point up and others down? (b) Features 2 and 6, and 3 and 5 are slightly offset in the left-to-right direction with respect to each other. Is this correct? If so, why?

3.10 (a) Create a sketch similar (but not identical) to Fig. 3.6a, with both topographic variations and two friction coefficients. Include a horizontal surface. Then sketch the corresponding friction loop below it, similar to how Fig. 3.6b is immediately below the topographic sketch and has the same x-scale. (b) During class, you and your fellow students can go to the board to sketch your topographic cross-sections with two friction coefficients. Then all students can go to other students' sketches and draw friction loops.

3.11 In Fig. 3.6c, four force components (labeled 1–4) are shown. (a) How are their magnitudes related to the normal force N and friction force μN? (b) If the direction of the tip is instead down the slope, which forces and components change? Describe the changes. (c) Rewrite Eqs. 3.4 to represent moving down the slope.

3.12 Take the limits of Eqs. 3.7 for the cases of (a) $\omega = 0$, (b) $\omega^2 = k_c/m$, and (c) $\omega = \infty$.

3.13 (a) What are the units of the $\tan\theta$ equation in Eqs. 3.7? (b) What are the units of b_c? (c) Show that the units of the left- and right-hand sides of the c_1 equation are consistent.

3.14 (a) If the total energy of an oscillating cantilever is 2.00×10^{-13} J, what is its maximum kinetic energy in Joules? In MeV? (b) If its effective mass is 11.3 ng, what is its maximum speed in cm/s? (c) If its spring constant is 40.0 N/m, what is its amplitude (magnitude of its turning points) in nm?

3.15 Make a table about AFM imaging modes of operation and label the columns Contact, Non-contact, Intermittent-contact, and Off-resonance. Label the rows Description, Pros, Cons, and Uses. Make comments in the empty squares.

3.16 Lab question. Assuming that you have already done some phase imaging, what causes the phase to be delayed or advanced?

3.17 *Challenge Problem.* (a) Solve Eqs. 3.3 and your answer to P. 3.11c for F_{lat+} and F_{lat-}. (b) Ensure that you can end up with Eqs. 3.6 from Eqs. 3.5.

Chapter 4
AFM design and calibration

Abstract This chapter covers AFM design and calibration. Design concerns the choices one makes in order to most easily obtain the desired data. Calibration concerns the steps to accurate quantitative results. We focus on two essential components, the probe and the scanner. The probe measures the changes in topography or force, and the scanner moves the probe over the sample, or, in some AFMs, the sample under the probe. Both components need calibration. The probe consists of the tip and cantilever. The cantilever's spring constant and deflection determine the tip-sample force. After providing background information, the advantages and disadvantages of design choices are treated, followed by summaries of the calibration methods available. The goals are for you to be able to choose your probes and scanners appropriately and to be able to calibrate them. These goals will, in turn, speed you towards obtaining good AFM data quickly and accurately.

4.1 Probe design and calibration

This section begins with a discussion of the typical materials and geometry of probes and the properties of cantilevers. Then we proceed to cantilever and tip design, followed by force and tip calibration. Force calibration means determining both the spring constant and the deflection of the cantilever, and tip calibration means establishing its size and shape.

4.1.1 Probe design

4.1.1.1 Materials and geometry

The probe chips, cantilever beams, and tips are usually made from silicon (Si) or silicon nitride (Si_3N_4), then usually coated with a reflective metal. The reflective coating increases the proportion of light that is bounced back towards the photodiode. The probes are made at facilities similiar to ones that make the processors in computers. These facilities are optimized for very specific kinds of materials, which is why commercially available cantilevers are usually fabricated from silicon or silicon nitride.

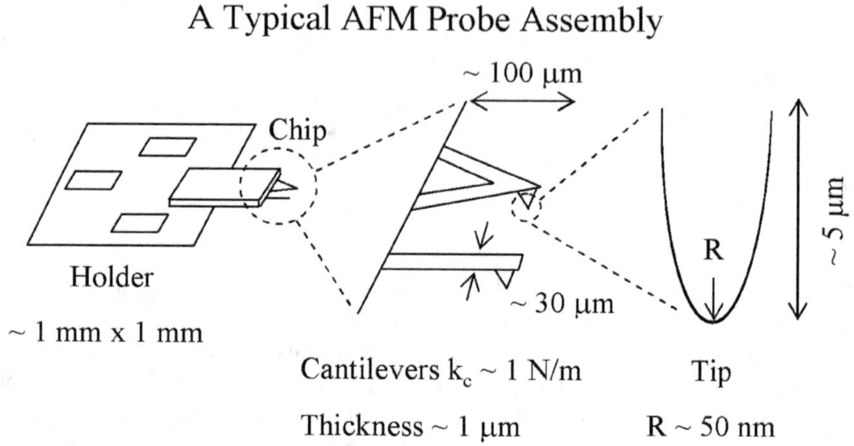

Fig. 4.1 Overview of a typical AFM probe assembly. On the left is a view of the holder and chip. In the center are shown two common forms of cantilever, triangular and rectangular. The part of the probe that touches the sample is the tip, seen on the right. Typical values for dimensions and spring constants are displayed.

Many cantilever beams have the appearance of miniature diving boards, as depicted in Fig. 4.1 (the same figure as Fig. 1.2, repeated here for your convenience). The cantilevers are typically 100–500 μm in length, 30–40 μm in width, and 0.2–2 μm in thickness; some are much smaller. Sometimes two such rectangles are made to connect so as to form a V-shaped or triangular cantilever, which was originally believed to help the cantilever resist torsional motion. Both the geometry and the properties of the material determine the two most important characteristics of the cantilever beams—the spring constant (Sect. 1.2.1) and the frequency at which the beam naturally vibrates. The natural frequency must be higher than the building's vibrations such that any low-frequency building vibrations do not cause the cantilever to oscillate (Sect. 2.3). This characteristic is achieved only if the cantilever (or any object) is small.

Tips come with different shapes. Integrated tips are made at the same time as the chips and cantilevers, in the processing facility; tips can also be added after processing. The shapes of the integrated tips are determined by the fabrication process and are limited by the properties of the material. Silicon tips are conical, narrow, and more brittle as compared to silicon nitride tips, which are pyramidal, broader, and more durable. For either a silicon or silicon nitride tip, the radius of curvature of its apex is usually in the range of 10 to 100 nm, with some suppliers offering tips for high-resolution imaging with radii less than 10 nm. After the initial fabrication, the shapes of individual tips can be modified by exposure to a focused electron or ion beam to make a very fine point, smaller than 10 nm. Alternatively, cylindrical molecules known as carbon nanotubes can be attached to tips. These molecules have very small end radii of curvature as well. And, if a spherical tip is desired, spheres of polystyrene or glass can be attached to the cantilever. They have radii in the micron range.

Tips come with different surfaces. Silicon forms a natural amorphous oxide, so the surface that contacts the sample is in fact SiO_x, where the x means that its exact composition is unknown. Both silicon oxide and silicon nitride may absorb water and other contaminants from the air. But the researcher is not limited to these two surfaces. The tips may be coated with various magnetic materials and metals. The metals can then be modified by the purposeful attachment of different molecules. If a carbon nanotube has been attached to a tip, the end of the nanotube can be chemically changed.

4.1.1.2 Properties of the cantilever

The two most important properties of the cantilever are its *spring constant* or *stiffness* and its natural frequency. The spring constant relates the deflection to force. Hooke's Law formalizes the relationship.

$$\mathbf{F} = -k\mathbf{d}, \tag{4.1}$$

where **F** stands for force (a vector), k for the spring constant or stiffness of the cantilever, and d its deflection (a vector). The word "deflection" is a reminder that for most AFM detection systems, the important aspect is not the absolute position of the tip end of the cantilever, but rather the change from its free equilibrium position. The significance of the minus sign is to show that the direction of the force of the cantilever on the tip is in the opposite direction to the deflection. Hooke's Law establishes the necessity of obtaining accurate values for the spring constant k and deflection **d** in order to have a trusted value for force. Often one sees the relationship among the magnitudes of the variables, $F = kd$. (Aside: If you see this written as $F = -kd$, the author is not paying attention to vector notation!)

Spring constants are calculated using torque analysis, where the torque acting on a differential volume, dv, of the cantilever (Fig. 4.2a, shaded) is equal to $\mathbf{r} \times \mathbf{F}$ (Fig. 4.2a, bold vectors). The total vector torque is found by integrating along the length of the cantilever. From each component of the torque, the force can be expressed as a constant times a deflection; the constant is the spring constant. The most commonly used one is the normal spring constant, k_{norm}, for motion normal to the plane of the cantilever. (It is so common that in other chapters of this book, a simpler notation is employed, k_c, for cantilever spring constant.) The next most important one is k_{tor}, for the torsional twisting of the cantilever, used in lateral force microscopy (Sect. 3.3). And the third one is k_{lat}, for lateral flexing within the plane

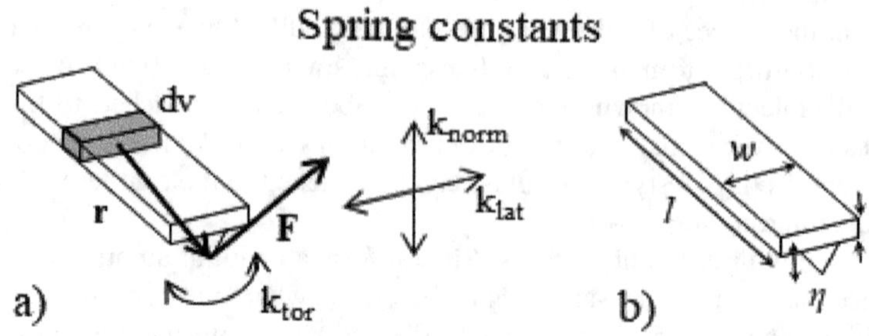

Fig. 4.2 (a) Spring constants are calculated by finding the torque $\mathbf{r} \times \mathbf{F}$ (bold vectors) acting on a differential volume dv (shaded), then integrating along the length of the cantilever. For each spatial component of the torque, a force is related to a deflection. The proportionalities are the spring constants in the normal, lateral, and torsional directions. In (b), the length l, width w, thickness τ, and tip height η are shown.

of the cantilever; most AFMs are not able to detect this motion. Figure 4.2a shows the directions of the three spring constants.

The spring constant k depends on the properties of the cantilever material as well as its geometry. For a rectangular beam of length l, width w, thickness τ, (Fig. 4.2b) volume density ρ, and elastic modulus E (the inherent stiffness of the material in N/m^2, Sect. 5.3.1), the normal spring constant is

$$k_{norm} = \frac{Ew\tau^3}{4l^3}. \tag{4.2}$$

The lateral spring constant has a similar form. Only the powers of the width and thickness are exchanged.

$$k_{lat} = \frac{Ew^3\tau}{4l^3}. \tag{4.3}$$

And as long as the cantilever's width is much greater than its thickness, the torsional spring constant is

$$k_{tor} = \frac{Gw\tau^3}{3l\eta^2}, \tag{4.4}$$

where the new variables are G, the shear modulus of the material in N/m^2 (its inherent resistance to twisting), and the distance from the tip to the axis around which the twisting occurs, η. The above are exact for when the plane of the cantilever is parallel to the plane of the sample's surface. Small corrections are necessary when the cantilever is inclined with respect to the sample, as it is in most AFMs, but need only be done if accuracy in force must be better than 10%.

Manufacturer's values for cantilever spring constants can be off by a factor of three! Why so much? The smallest dimension of the cantilever, its thickness, is difficult to control during its fabrication. As the spring constant depends on the cube of the thickness, even small variations in thickness have a big effect.

The other important property of a cantilever is its *natural frequency*, ν_k. One can use equations developed to predict the vibrations of mechanical structures in order to find the frequency at which a uniform rectangular beam naturally vibrates,

$$\nu_k = \frac{1}{2\pi}\frac{\tau}{l^2}\sqrt{\frac{35E}{33\rho}}. \tag{4.5}$$

A consequence of the analysis leading to this equation is that the effective mass of the cantilever is about a quarter of the volumetric mass; the mass near the fixed base of the cantilever does not contribute as much to its motion as mass near the free end of the cantilever.

As a last note on cantilever properties, Eq. 4.5 refers to vibrations in the normal direction. Most cantilevers are constructed such that the natural frequencies in the lateral direction and for twisting are very high. They are not usually detected.

4.1.1.3 Cantilever design

There are always optimizations to be made when designing any device, even one so simple as a rectangular cantilever. The lowest natural frequency (Eq. 4.5) of the cantilever should be great enough such that it will not be sensitive to building, pump, or compressor vibrations, which are typically in the range of 10–1000 Hz. The tip radius should be small for good lateral resolution, yet this means large force per unit area (pressure), which could damage the tip or sample. Thus the spring constant should be low, so that the pressures remain low. In other words, the probe designer wants to fabricate a cantilever with a high ratio of frequency to spring constant, ν_k/k_{norm}. From the expressions for ν_k and k_{norm} (Eqs. 4.5 and 4.2, respectively),

$$\frac{\nu_k}{k_{norm}} = \frac{2}{\pi}\sqrt{\frac{35}{33\rho E}\frac{l}{w\tau^2}}. \tag{4.6}$$

The variables before and underneath the square root are constants or are material-dependent parameters. Once the choice of material (e.g. silicon or silicon nitride) has been made, they are fixed. The width w must be large enough to reflect most of the light that shines upon it, so even though reducing this value would help the design parameter ν_k/k_{norm}, it does not usually vary from the range of 30–40 μm unless special optics have been used to reduce the width of the illuminating beam. Thus the designer varies the ratio of the length of the cantilever to its thickness squared, l/τ^2. A long, thin cantilever will be best for not being susceptible to building vibrations and for a low spring constant. Yet, it is possible to have such a low spring constant that the cantilever vibrates due to thermal energy.

Thermal energy is the energy that an object has even if it otherwise totally isolated from its environment and nominally at rest. Thermal energy is small, much less than the energy in a single photon of visible light. Thermal energy causes the cantilever to vibrate a tiny bit, such that its total energy is cycled between small kinetic motions and the potential energy of the "spring." Incredibly, these motions can be detected in an AFM. The following equation relates temperature (Υ) in Kelvin to the spring constant and deflection, making use of the Boltzmann constant k_B, with k_B equal to 1.3806×10^{-23} J/K.

$$\frac{1}{2}k_B\Upsilon = \frac{1}{2}k\left\langle d^2\right\rangle, \tag{4.7}$$

where $\left\langle d^2\right\rangle$ is the mean-square value of the deflection (Eq. 2.2). You will calculate some root-mean-square deflections in P. 4.10.

Stiff cantilevers can apply a larger range of forces. Compliant ones offer better force resolution. The choice is between range and resolution. The spring constants of commercially available cantilevers cover five orders of magnitude, from approximately 0.001 to 100 N/m. For perspective, 100 N/m is the stiffness of a spring on the inside of a ball-point pen. The children's toy called Slinky is near 1 N/m, and the bonds between atoms are on the order of 1–10 N/m. Usually, lower spring constants are used in contact mode, moderate ones in non-contact mode, and high ones in intermittent-contact mode (Sect. 3.4). The natural frequencies of commercially available cantilevers range between several kilohertz to megahertz.

> **AFM Lesson 16 at `youtube.com\AtomicForceMicro`**
>
> AFM Lesson 16 considers cantilever design and selection.

4.1.1.4 Tip design

The tip is the part of the probe that contacts the sample. Whereas your choice of a cantilever is an issue of "range versus resolution," the issue for tips is "robustness versus resolution." For cantilevers, range and resolution refer to force. For tips, robustness refers to its durability to the pressures that act upon them, and resolution refers to the lateral resolution of an image. Durable, robust tips are generally bigger and can be used longer before their size and shape change. But they are like big, fat fingers attempting delicate work. Small tips make small contacts with the sample; thus tiny features can be imaged, although high forces must be avoided in order that they maintain their shape.

One can also choose a tip with a very well known geometry, for example a sphere. Spherical tips also hold the advantage that the projection of the contact they make with the sample onto a plane will remain circular, regardless of the angle of repose of the cantilever. Figure 4.3 shows how the contact can become distorted with a conical

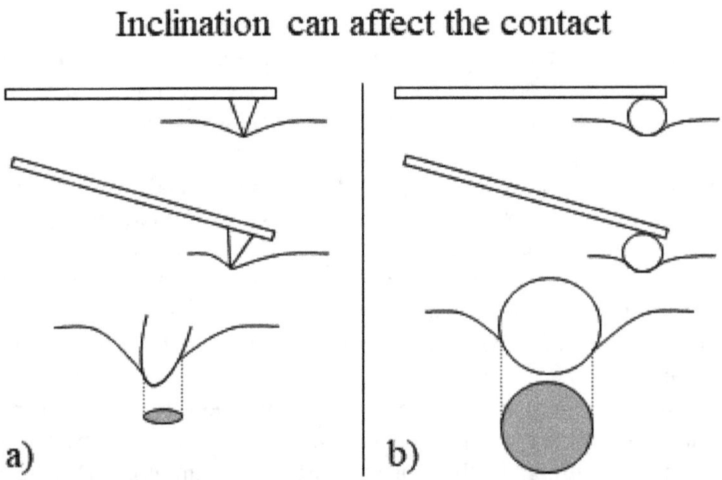

Fig. 4.3 (a) Conical and (b) spherical tips contacting horizontal smooth surfaces. The top sketches show the cantilever and sample in the same plane, the middle ones depict the cantilever inclined with respect to the sample, and the bottom ones are close-ups of the contact. The shaded shapes on the very bottom represent the shapes of the contacts as viewed from above. A conical tip on an inclined cantilever forms an elliptical contact with the horizontal sample, making analysis more difficult. In comparison, a spherical tip forms circular contacts independent of the cantilever's angle of repose.

tip on an inclined cantilever. The disadvantages to spheres are that they are not always smooth and they usually have radii in the range of 1–10 μm. In comparison, integrated tips have nanometer-scale radii.

4.1.2 Force calibration

Because force is the product of the spring constant and the deflection, both must be known in order to have a quantitative value for force.

4.1.2.1 Calibration of spring constants

Many different approaches to the calibration of normal spring constants have appeared in the literature, too many to review individually here. However, the methods can be grouped into four different types: theoretical, geometric, thermal, and loading.

Theoretical methods involve detailed calculations of the expected stiffness of the cantilever based on its geometry and materials properties. The geometry might be established with scanning electron microscopy or the dimensions might merely be read from the manufacturer's specification sheet. The materials properties are usually taken from a reference book. The calculations are precise enough to take into account the properties and thickness of the reflective coating. No experimental measurements are taken, however.

Geometric methods, similarly to the theoretical methods, rely on some of the dimensions of the cantilever and the properties of the material(s) out of which the cantilever is fabricated. In contrast with the theoretical methods, the geometric methods require at least one other experimental parameter to be measured.

Loading methods involve the application of a known force to the cantilever and observing its deflection. The force can be applied to the cantilever in a variety of ways: by a calibrated instrument known as a nanoindenter, by adding known masses to the cantilever's free end, by covering the cantilever with a known thickness of coating, and by pushing a cantilever of unknown stiffness against one of known stiffness. The US National Institute of Standards and Technology uses this approach, but it requires significant infrastructure.

Thermal methods make use of the idea that temperature is equivalent to energy. Because of temperature, an object vibrates, and these vibrations can be seen in noise power spectra, which were discussed in Sect. 2.3.1. A power spectrum of the cantilever's free motions is collected. The contributions to the spectrum from 1/f noise, white noise, and from the cantilever itself are established with a curve-fitting routine. Then the y-values of the cantilever's contribution to the power spectrum are added to find the mean-square value of the cantilever's thermal fluctuations, $\langle d^2 \rangle$. From Eq. 4.7, the spring constant can be found. The accuracy in the values for spring constants from these methods can be smaller than 10%. Many modern commercially available AFMs now come equipped with the necessary hardware and software for this approach. The precaution the researcher must take is to ensure that the energy available to the cantilever is only the thermal energy of the laboratory. The intensity of the illumination must be low enough not to significantly raise the temperature of the cantilever, and mechanical and acoustical excitation must be lower than thermal excitation.

If your AFM is not configured to provide a spring constant for your cantilever, the quickest calibration method for the stiffness is a geometric one. You will use the manufacturer's values for the width, length, density, and elastic modulus. You will measure the resonance frequency, which is proportional to the thickness of the cantilever. Thickness is important because the normal spring constant depends on its cube (Eq. 4.3). Unfortunately, thickness is the most challenging dimension to control during probe fabrication. It even varies from probe to probe during the same fabrication run. And it is too small to be measured accurately with an optical microscope. The expression for the resonance frequency, Eq. 4.5, is solved for the thickness τ and inserted into the definition of the normal spring constant, Eq. 4.2. For a rectangular cantilever, this process yields

$$k_{norm} = 2w \frac{(\pi \nu_k l)^3}{E^{1/2}} \left(\frac{33\rho}{35} \right)^{3/2},$$
$$= 56.77w (\nu_k l)^3 \sqrt{\frac{\rho^3}{E}}. \quad (4.8)$$

For triangular cantilevers, a reasonable approximation for the stiffness is to multiply the right-hand side by a factor of two.

There have been dozens of different methods for calibrating the spring constant, yet just two methods are commonly used for the deflection.

4.1.2.2 Deflection calibration

The change in voltage of the photodiode must be converted into the deflection, that is, the distance the cantilever has bent from its free equilibrium position. Typically this is accomplished by placing the tip in contact with a stiff surface and ramping the scanner movement such that the cantilever bends backwards. If the surface is truly much stiffer than the cantilever, the cantilever will move as much as the sample, without indenting it. The change in voltage from the photodiode for a given (known) movement of the scanner yields the conversion factor necessary to calculate deflection from the photodiode signal. (As will be explained in Sect. 4.2, there can be challenges associated with knowing the scanner's movements. Still, scanners can be calibrated by a number of methods.) The process by which the conversion factor is found is described in more detail in Sect. 5.2.

Alternatively, one can first use, say, a geometric method for the spring constant, then do a thermal measurement to calculate $\langle d^2 \rangle$ from Eq. 4.7. The ratio between the $\langle d^2 \rangle$ and the mean-square voltage from the photodiode is the deflection calibration value. The advantage of this approach is that the tip need not touch a stiff sample before examining the sample of interest.

Every time that the laser light is readjusted, the deflection sensitivity should be calibrated again; it is dependent on the position of the light spot along the cantilever and how much light is reflected to the photodiode.

4.1.3 Tip calibration

It is important to know the tip's size and shape in order to estimate the expected lateral resolution, to monitor wear, and for force-curve experiments (Chap. 6) that investigate the nanoscale properties of materials. Tip calibration methods fall into four categories; the first is accomplished with electron microscopes, and the other three are done with AFMs. Each has its own advantages and disadvantages.

By placing the probe into a *calibrated electron microscope*, the tip can be imaged and its size and shape easily seen. This is a very direct method. However, since electrons are used to produce the images, the probe must be conducting. Most silicon cantilevers are modified during manufacturing to be conducting, but silicon nitride cantilevers and the oxide layer on silicon cantilevers are not. A thin coating of metal can be grown on the cantilever, but this process incurs the disadvantages of modifying the cantilever from its original state and adding to the tip's radius of curvature. Moreover, if the relatively expensive electron microscope is not easily accessible, this approach can add a significant amount of time and possible expense to tip calibration.

By subjecting an AFM image to a software protocol that looks for the smallest radii of curvature at each height level of the tip, the size and shape of the tip can be reconstructed from the image. This mathematically advanced process is known as *tip reconstruction*. The advantage to tip reconstruction is that it can be done from the image that one wants to acquire for other reasons. A disadvantage is that sometimes there are not enough features or rugosity on the sample to complete a full reconstruction. Sometimes this form of calibration is done with a calibration sample that has sharp, robust edges.

By indenting a soft plastic material such as indium with a large force at one point on the surface (*plastic indentation*), a hole is left that can be later imaged at a small force. The image of the hole reflects the tip size and shape. Usually an array of indentations is made; each one is made a different depth such that a three-dimensional model of the tip can be replicated. The disadvantage here is that most AFMs are made such that the tip slides along the surface during the indentation process (Sect. 5.2.3), distorting the hole along the sliding axis.

Finally, if the tip passes over a feature that is significantly smaller than the tip itself, the image that results is an inverted image of the tip. Seen from one point of view, this is the *tip-imaging* artifact (Sect. 2.2.1.7). Performed purposefully, it reveals the tip size and shape in a few minutes. Samples made with sharp features can be purchased from probe suppliers. However, small features wear easily, and the manufacturer's stated feature size may evolve with each pass of the tip. The features can be steps, spikes, or holes. They and the paths that a pyramidal or conical tip would follow are displayed in Fig. 4.4. Steps are commonly used for large radii ($R \gtrsim 1\ \mu m$), spikes for intermediate radii ($1\ \mu m \gtrsim R \gtrsim 100\ nm$), and holes for small radii ($R \lesssim 100\ nm$). Steps image just one side of the tip at a time; spikes wear easily; holes clog with dust and debris. Another disadvantage to spikes, or delta functions, is that if their end radii are of the same order of magnitude of the tip, one cannot know if the observed radius in the image is due to the tip or the spike.

The last aspect of designing and calibrating an AFM concerns the scanner, which is the topic of the next section.

AFM Lesson 17 at youtube.com\AtomicForceMicro

AFM Lesson 17 describes options for probe calibration.

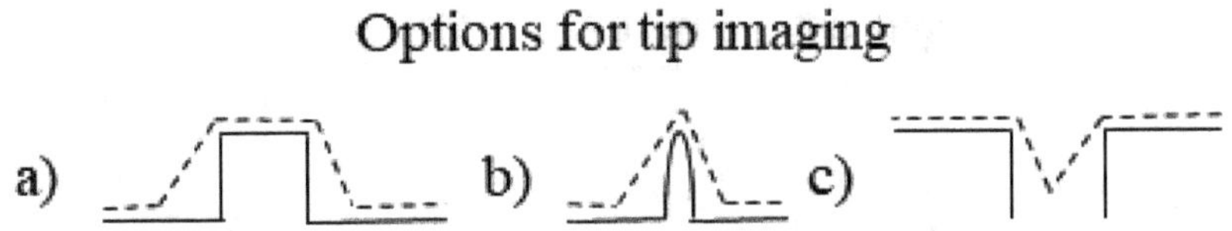

Fig. 4.4 Three common samples used for imaging the tip in order to obtain an inverted image of the tip: (a) steps, (b) spikes, and (c) holes. The dashed lines show the path that a conical or pyramidal tip would follow. The inclined dashed lines are not symmetric about the vertical axis because the tip twists in response to contact with the sample.

4.2 Scanner design and calibration

Different AFM manufacturers choose different scanner configurations, different types of scanners and different ways of compensating for their problems. An overview of these choices is given in this section, as well as guidance for calibrating them.

4.2.1 An AFM's scanner configuration

In order to image, an AFM probe's tip must move relative to the sample. There are three common choices to do so. Traditionally, a small sample (on the order of a square centimeter) was placed on the scanner and moved in all three (x-y-z) directions, as depicted schematically in Fig. 4.5a. This choice is still used when exciting the tip-sample interface with light in order to map, for example, infrared absorption, such that the tip and excitation remain stationary.

The scanning can also be done completely from the tip side (Fig. 4.5b); this configuration is chosen for large samples, in that the AFM head can be placed on top of or suspended above a sample, such as a 300-mm diameter silicon wafer. In some cases, the sample is placed on a motorized stage, such that the sample can be moved over centimeters and the AFM images collected from anywhere on the sample.

As detailed below in Sect. 4.2.3, scanners have their imperfections, and precise movements can be more easily achieved if the x-y-z motions are decoupled. In this case, a scanner below the sample controls the horizontal motions (x-y) and a z-scanner controls the vertical motion of the probe (Fig. 4.5c). This choice allows the z-scanner to move quickly due to the relatively small and consistent mass of the probe, which lowers the chance of inadvertent tip crashes. In this case, some x-y-scanners accommodate samples up to the size of approximately a microscope slide; others are designed for larger samples such as silicon wafers. Similarly to the tip-scanning situation, the x-y scanners are sometimes mounted on a motorized stage. A caution for scanning large, more massive samples, is that x-y scan frequencies generally do not exceed 1 Hz, meaning that images can take some minutes.

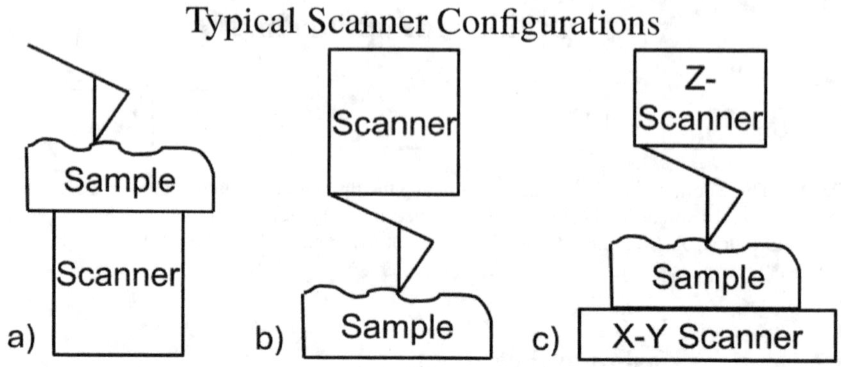

Fig. 4.5 Typical AFM scanner configurations. (a) Traditional scanner beneath the sample; (b) tip scanning; (c) decoupled scanning, with horizontal (x-y) movement of the sample and vertical (z) movement for the probe. There is no "right" configuration; the best choice depends on the size of samples and the intended studies.

4.2.2 Operational principles of common scanners

The most common types of scanners are based on voice coils or piezoelectric actuators. Voice coils use the same technology that drives speakers or earbuds. The coils are connected to the speaker's or earbud's diaphragm and are mounted on a light spring. Current passing through the coil of wire generates a magnetic field inside the coil. The coil's magnetic field interacts with a permanent, fixed magnet mounted next to the coil, creating a force between them. The force moves the diaphragm back and forth, creating the acoustic wave that you hear from the speaker or earbud. For AFM, instead of voice coils being connected to diaphragms, they drive the scanning.

Scanners based on piezoelectric materials have been implemented in several different geometries: stacks, tubes, and flexure stages, where the piezoelectric materials drive the flexing of a sample stage. Independent of the geometry, they operate on the same principle. A ceramic composed of grains of polar molecules, typically lead zirconate titanate, is sandwiched between electrodes. Each grain has a dipole moment. The dipole moments are oriented mostly in the same direction. A voltage difference is applied across the electrodes, creating an electric field across the material. The force created by the electric field on the dipoles causes them to rotate, hence expanding or shrinking the material between the electrodes, thereby driving the scanning.

4.2.3 What are typical scanner problems?

Voice coils are inexpensive but slow. Lenz's Law, which states that a changing magnetic field creates a current that opposes the current that creates the magnetic field, limits their scanning speed. Furthermore, in order to hold a non-zero position, current must run through the coil, which causes heating and thermal expansion of components and subsequently drift of the components relative to each other. Voice coils are also limited in the amount of force they can generate, thus they have been implemented only in tip-scanning configurations so that the typically more massive samples can be stationary. However, they operate with lower hysteresis and creep than piezoelectric scanners.

Turning now to piezoelectric scanners, the ceramics from which they are made are stiff. As a result, the scanners have high resonance frequencies, enabling them to move quickly and making them less sensitive to building vibrations (refer to Sect. 2.4). However, because the dipoles of the many grains in the ceramic material are not perfectly aligned and therefore respond slightly differently to the applied scanning voltage, they are subject to various nonlinearities that will be described briefly below.

To appreciate the extent to which piezoelectric scanners behave nonideally, inspect Fig. 4.6a. In it, a uniform grid with a repeating pattern of period 10 μm in both the

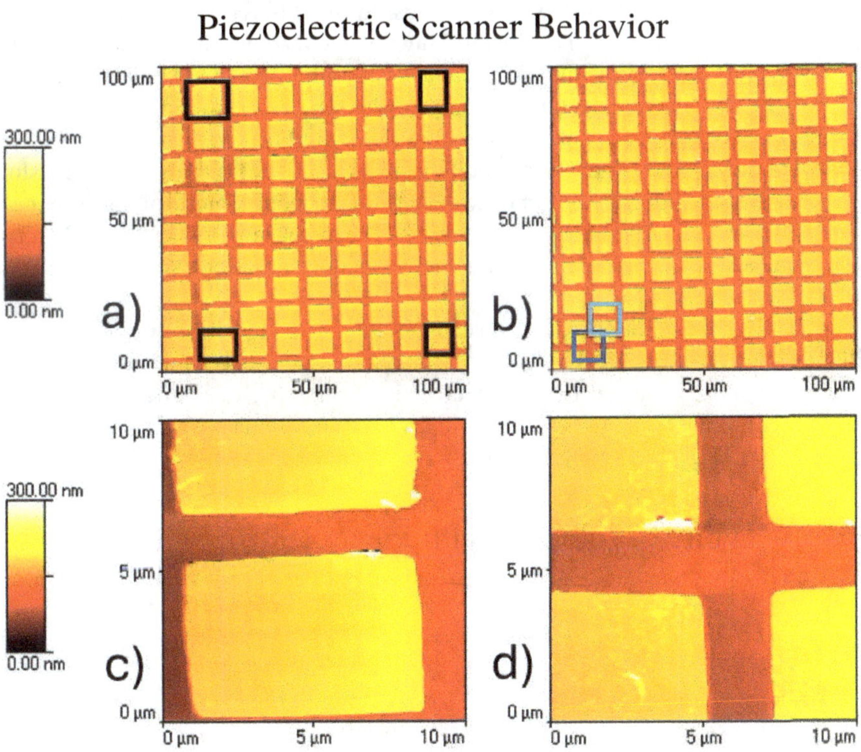

Fig. 4.6 Piezoelectric scanners do not behave ideally unless they are corrected by means of an independent feedback loop. (a) Uncorrected large scan of a 10 μm-period grid. Four "squares" are highlighted to show how nonuniform they are; (b) corrected version; (c) without correction, an attempt to zoom into the area highlighted in light blue in (b); (d) with correction, the zoom into the intended area (dark blue) is almost completely successful.

x- and y-directions has been imaged, but instead of squares of uniform size, we see both rectangles and squares of varying size. Four "squares" have been highlighted to guide your eyes. Figure 4.6b is the corrected version. Various techniques for correcting nonlinearities are described in the next section; they all consist of an independent feedback loop that compares the actual position of the scanner with its intended position.

Zooming into a small feature near the edge of a large scan range can be difficult. Without correcting the piezo motion, the intended area of a small scan, shown in Fig. 4.6b as a dark blue square, instead scans the area of the light blue square, yielding Fig. 4.6c. With correction, the scanner images the area in Fig. 4.6d. While the "plus sign" formed by the trenches between the squares should have been exactly in the middle of Fig. 4.6d, the scanner did behave much better with correction in Fig. 4.6d than without correction in Fig. 4.6c.

As mentioned above, the nonlinearities of piezoelectric scanners originate from the structure of the dipole grains and sometimes the geometry of the scanner. Common problems are hysteresis, creep, aging, and cross-coupling.

• *Hysteresis* concerns the repeatability of the motion forward and back or up and down. Figure 4.7a shows how scanner position is different for increasing voltage (arrows up and to the right) and decreasing voltage (arrows down and to the left).

• *Creep* describes how the piezoscanner will initially respond quickly and nearly fully to the applied voltage, but then only slowly achieve the full expected range over a time scale of seconds, as depicted schematically in Fig. 4.7b.

• *Aging* indicates that the orientation of the dipoles of the grains will slowly become more random over time, lowering the responsiveness of the scanner to voltage (Fig. 4.7)c. Thus if a piezoscanner has been unused for a few weeks it should be exercised at a high scan range overnight and its calibration (Sect. 4.2.5) checked in the morning.

• *Cross-coupling* applies to tube scanners configured to deliver motion in all three dimensions. As the x- and y- positions increase, the tube assumes a parabolic-like shape, and more voltage is needed to move the scanner in z to get the same amount of vertical movement as for when the x- and y-positions remain small. In other words, the z-response goes down (Fig. 4.7d).

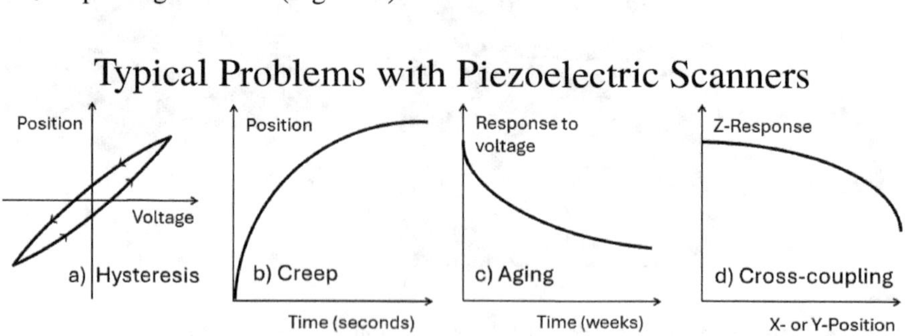

Fig. 4.7 Typical problems with piezoelectric scanners. (a) hysteresis, when applying and removing voltage results in different positions; (b) creep, when the final position is achieved over a short time instead of instantly; (c) aging, when the response decreases with no use over weeks; and (d) cross-coupling, when the z-reponse decreases at high x- and y-positions.

4.2.4 How do we correct for the problems?

The nonlinearities associated with scanners can be corrected by implementing an additional feedback loop that compares the desired position of the scanner with its actual position, then brings the scanner to its desired position. This is called *closed-loop scanning*. This approach is integrated into many AFMs with larger maximum scan ranges. Figure 4.6 is an example in that a large scan range of 100 μm was used; scanner problems are less evident at smaller scan ranges.

The additional feedback loop adds noise to the image. Users often turn off the closed-loop scanning for small scan ranges and when the features of interest are small. The additional feedback loop makes the AFM more expensive as well. Part of the expense is in the electronics of the feedback loop and part is in the sensor that detects the actual position of the scanner.

Three common sensors for closed-loop scanning are strain gauges, capacitors, and linear variable differential transformers. These are described forthwith.

• *Strain gauges* are based on the principle that a metal's resistance changes with its geometry, and the geometry changes with the strain that it is under. A metal is printed on a flexible film in a raster pattern, and the metal and film are glued to the scanner. As the scanner moves, the strain gauge expands, compresses, or flexes, and its resistance changes. The resistance provides the measure of the scanner's position. Strain gauges are inexpensive, but tend to be less durable with repeated use because of fatigue in the metal and the long-term adhesion of the glue to the scanner. Moreover, since a metal's resistance is dependent on temperature, they are sensitive to temperature.

• *Capacitors* are typically set up in parallel-plate configuration, such that one plate moves with the scanner and one plate is fixed on the AFM. As the spacing between the plates changes, the capacitance changes, which is the measure of the scanner's position. The necessary electronic circuitry is more expensive than that for strain gauges, but capacitive detection exhibits very good long-term stability and temperature independence.

• *Linear variable differential transformers (LVDTs)* transform mechanical motion into a voltage difference. They are configured as a pole within a tube, with the magnetizable pole connected to the scanner and the tube consisting of three equally spaced coils of wire. The AC current running through the center coil keeps the pole's magnetic field fluctuating, which by Faraday's Law is necessary for creating a voltage in a nearby conductor. The circuit is set up such that if the pole moves, a linear difference voltage is created between the outer coils if the pole is off center. Like capacitors, LVDTs are used when long-term stability and temperature independence are worth the additional expensive. However, they typically require more space than parallel-plate capacitors.

Even with closed-loop detection, scanners must still be calibrated such that they move the expected distances.

4.2.5 How do we calibrate the scanner?

AFM users should be confident that the scan range that they input into their scan parameters is correct: calibration is essential. To calibrate, a calibration sample is imaged and its dimensions read from the image. If the dimensions are not what they should be, the user can make a software adjustment such that the amount of voltage applied to the scanner is modified to return the correct dimensions the next time the sample is imaged. The calibration factor changes slowly with time and with the usage of the scanner, as described in Sect. 4.2.3. When highly accurate dimensions are important to your research, checking the calibration once a week is recommended. Generally, a few times a year is sufficient.

There are two classes of calibration samples: calibration references and calibration standards. The *calibration references* might offer a percent standard deviation in their dimensions of 3% and can be purchased for a few hundred US dollars (as of this writing). The *calibration standards* are "traceable," meaning that their design and processing originated at a national standards laboratory, such as the US National Institute of Standards and Technology. These standards are more expensive, generally over $1000, but with percent standard deviations for the dimensions less than 1%. Depending on the need for accuracy in your laboratory, they may or may not be worth the extra expense.

When you purchase calibration samples—be they references or standards—be aware that the quoted standard deviations for the dimensions in x, y, and z might be different. For AFMs with large scan ranges, consider buying calibration samples with both large and small periods, for example, a 10-μm period for large x-y scan ranges, and a 300-nm period for small scan ranges. Similarly, for the z-direction, consider buying steps of known heights in the micron range, along with steps in the nanometer range.

If your AFM has good vertical resolution, consider buying graphite, a two-dimensional layered compound of carbon. Graphite holds the advantage that it can be easily exfoliated to produce a clean surface each time you run a calibration scan, yielding an atomically flat surface with steps at the interfaces of grains, corresponding to an integral number of graphite layers. The distance between single layers is well established to be 0.335 nm. (The carbon atoms within a layer are 0.142 nm apart, a size scale that is challenging for most AFMs.) The disadvantage of graphite is that it is not easy to discern exactly how many layers constitute a given step.

If your funds are limited for buying calibration samples, an inexpensive alternative for an x-y reference is a diffraction grating, often used in undergraduate optics laboratories. An alternative to calibrate the z-response of your scanner is to use the following interference effect. This approach assumes that you know the angle of repose of the cantilever and the wavelength of the AFM's laser very well. the manufacturer should be able to tell you the nominal values, although double-checking the values by independent means (e.g., micrometers for angles, diffraction gratings or spectrum analyzers for wavelength) is recommended. By placing the laser half-on and half-off a cantilever above a reflective sample, a sinusoidal interference pattern can usually be discerned as the scanner is brought towards and away from the sample (Sect. 5.2.3.7). In P. 4.19, you will derive the appropriate equation for the path difference between the light reflected from the cantilever and the light reflected from the sample. Then, knowing the wavelength of your laser, you will be able to confirm the scanner's z-movement.

AFM Lesson 18 at `youtube.com\AtomicForceMicro`

AFM Lesson 18 reviews scanner considerations and calibration.

In **Instrument Lab 4**, you explore options for probe and scanner calibration.

This concludes Part I on instrumentation. Part II concerns interpretation, emphasizing force-curve interpretation.

4.3 Learning objectives, further reading, and problems

! Learning objectives for Chapter 4

From the material in Chapter 4, the reader should now be able to:

4.1 Distinguish among the parts of a probe assembly.
4.2 Understand why certain materials and geometries are used to construct probes.
4.3 Discuss the phrases "range versus resolution" and "robustness versus resolution."
4.4 Describe the various ways in which tips and cantilevers are calibrated.
4.5 Describe the various ways in which deflection and scanners are calibrated.
4.6 Be familiar with common problems of scanners.
4.7 Explain how the scanner problems can be corrected.

Further Reading

✓**History.** Albrecht, T. R., Akamine, S., Carver, T. E., & Quate, C. F. (1990). Microfabrication of cantilever styli for the atomic force microscope. Journal of Vacuum Science & Technology A: Vacuum, Surfaces, and Films, 8(4), 3386–3396.

✓**Tutorial.** Griffith, J. E., & Grigg, D. A. (1993). Dimensional metrology with scanning probe microscopes. Journal of Applied Physics, 74(9), R83–R109.

✓**History.** Burnham, N. A., Chen, X., Hodges, C. S., Matei, G. A., Thoreson, E. J., Roberts, C. J., ... & Tendler, S. J. B. (2002). Comparison of calibration methods for atomic-force microscopy cantilevers. Nanotechnology, 14(1), 1.

✓**Tutorial.** Dzedzickis, A., Rožėnė, J., Bučinskas, V., Viržonis, D., & Morkvėnaitė-Vilkončienė, I. (2023). Characteristics and Functionality of Cantilevers and Scanners in Atomic Force Microscopy. Materials, 16(19), 6379.

? Problems for Chapter 4

4.1 If the *length* of a rectangular cantilever increases by a factor of two, by what factor does (a) the normal spring constant change? (b) the lateral spring constant? (c) the torsional spring constant? and (d) the resonance frequency?

4.2 If the *thickness* of a rectangular cantilever increases by a factor of two, by what factor does (a) the normal spring constant change? (b) the lateral spring constant? (c) the torsional spring constant? (d) and the resonance frequency?

4.3 A cantilever has a width 15 times that of its thickness and a length 10 times its width. (a) What is the ratio of its lateral to normal spring constants? (b) The ratio of its torsional to normal spring constants? Assume that the shear modulus is 1/3 that of its elastic modulus and the tip height is 5 times its thickness.

4.4 "Effective mass" equals the spring constant divided by the square of the radial frequency, $k_{norm} = m\omega^2$, where the radial frequency $\omega = 2\pi\nu$. (a) With the help of Eqs. 4.2 and 4.5, calculate the effective mass of a rectangular cantilever. (b) Calculate the volumetric mass (m_{vol}) by multiplying the volume density ρ of the rectangular cantilever times its volume. (c) Why are m and m_{vol} different?

4.5 For a summer internship, you are working at a probe manufacturing design center. Your boss asks you to design a probe for a customer who wants to do friction studies of high-friction materials with a probe that has a high torsional spring constant and high natural frequency. (a) You want to describe to the customer the characteristics of an appropriate probe: short or long, wide or narrow, thick or thin, short or tall tip. (b) Justify your proposed design with calculation(s) based on Eqs. 4.3–4.8. (c) Is your design consistent with a low normal spring constant?

4.6 The customer was pleased with your probes that you designed in the previous question. The new application area is fast scanning in contact mode, where the principle design parameters are low normal spring constant and high natural frequency. (a) You want to describe to the customer the characteristics of the cantilever: short or long, wide or narrow, thick or thin. (b) Justify your proposed design with calculation(s) based on Eqs. 4.3–4.8. (c) Does the tip height matter?

4.7 Continuing the same scenario as in the previous two questions, now a customer wants a low spring constant and high damping when the cantilever is near the sample surface. (High damping means that oscillations—intended or not—soon die out.) The damping coefficient for a long rectangular beam near a parallel surface equals

$\varsigma w^3 l/\eta^3$, where ς is the viscosity of the medium, e.g., air or water, and where the other parameters have been previously defined. (a) You want to describe to the customer the characteristics of the probe: short or long, wide or narrow, thick or thin, tall or short tip. (b) Justify your proposed design with calculation(s) based on Eqs. 4.3–4.8. (c) Would these characteristics be consistent with a high natural frequency?

4.8 A new customer wants different probe properties. State the desired properties (your choice). (a) Combine at least two equations in Sect. 4.1.1 for your design parameter, as was done for Eq. 4.6. (b) Describe the dimensions in words, i.e. long or short, wide or narrow, thick or thin, and tall or short tip.

4.9 What would you like to study with AFM? If you are not sure, ask a classmate or your instructor. Look online for a probe that might meet the requirements of the study. Choose one. Send the link to your instructor. Be prepared to explain to the class why you chose it.

4.10 Calculate the root-mean-square deflection, $\sqrt{\langle d^2 \rangle}$, due to thermal energy for cantilevers of stiffness 0.01 and 100 N/m, for both (a) room temperature and (b) the boiling point of liquid helium, 4K. Report your answers in picometers and femtometers, as appropriate. For perspective, the diameter of a hydrogen atom is 120 pm.

4.11 Prove that Eqs. (a) 4.2, (b) 4.5, and (c) 4.8 are dimensionally correct, that is, the units of the variables on the left- and right-hand sides are the same.

4.12 (a) In Fig. 4.4, which way is the tip moving? (b) Sketch how Fig. 4.4 would look if the tip were scanning in the opposite direction.

4.13 Go online and find at least one AFM manufacturer for each of the scanner configurations in Fig. 4.5. Send the links to your instructor, and be ready to present the information in class.

4.14 The resistance R of a conductor is dependent on its resistivity ρ, path length (total length of the raster pattern) L, and cross-sectional area A by $R = \rho L/A$. (a) If a strain gauge is made of a conductor with $\rho = 5.00 \times 10^{-7}\Omega\text{m}$, $L = 1.00$ cm, and $A = 50.0\,\mu\text{m}^2$, what is its resistance R? (b) If the strain gauge is stretched such that its path length becomes longer by 10 µm with negligible change in area, what is the new value of R?

4.15 The capacitance C of a parallel-plate capacitor depends on the permittivity of free space ε_0, the area A of the plates, and the distance d between the plates, $C = \varepsilon_0 A/d$. (a) Given that $\varepsilon_0 = 8.85 \times 10^{-12}$ C^2/Nm^2, show that the equation for capacitance is dimensionally correct, that is, the units on both sides of the equation are consistent. The unit of capacitance is Farads, equal to Coulombs per Volt, and a Volt is a Joule per Coulomb. (b) If $A = 1$ cm^2 and $d = 1$ mm, what is the capacitance C in picoFarads? (c) Sketch the capacitance for constant area as a function of d. Is it linear?

4.16 Voice coils and LVDTs function using magnetic fields created by currents within wires that have been formed into coils. The equation for magnetic field B within a long coil of N turns and end-to-end length L carrying current I is $\mu_0 NI/L$, where μ_0 is the magnetic constant, $4\pi \times 10^{-7}$ N/A^2 (Newtons per Ampere-squared). (a) Show that the equation for B is dimensionally correct, that is, the units on both sides of the equation are the same. Magnetic field is usually measured in Tesla. One Tesla (T) is one Newton per Ampere-meter. (b) For a wire carrying 10.0 mA of current that is wound into a coil of 100 turns over an end-to-end length of 1.00 cm, what is the magnetic field inside the coil in microTesla? (For comparison, the magnetic field of the Earth in most locations is about 50 μT.)

4.17 Lab question. Assuming that you have now spent at least one laboratory session on performing AFM calibration, suggest a plot for a Calibration Horror Movie and connect it to the content of this chapter. (Hint: What could be the societal consequences of improper calibration?)

4.18 *Challenge Problem.* Pretend that it is easy to measure the thickness of rectangular cantilevers and hard to measure their length. (a) Combine Eqs. 4.2 and 4.5 to find an expression for k_{norm} that does not include length. (b) Show that your answer is dimensionally correct.

4.19 *Challenge Problem.* A cantilever is positioned with an angle of repose Θ above a reflective sample. You place the laser half-on and half-off the cantilever, such that there is a path difference between the two halves of the laser beam. (While the sample is nominally horizontal, most are rough enough such that some reflected light from the sample reaches the photodiode.) As the scanner changes its z-position by Δz, a sinusoidal interference pattern is observed in the photodiode signal due to the difference in path lengths (Fig. 5.8b) of the two halves of the laser beam. (a) What is the difference in path length? (b) If you know the wavelength λ of the laser, what is the period of the sinusoid?

Part II
Interpretation

Chapter 5
Force-curve mechanics

Abstract An AFM image can be likened to moving your finger over a surface in a darkened room. You form a picture of the surface in your mind as your finger runs back and forth. Imaging is the original purpose and function of atomic force microscopy. Yet an important additional way of using AFM is to stop scanning and momentarily press the tip and sample together by ramping the scanner back and forth vertically, just as if you were to push into and then withdraw your finger from a material to test its properties. A compliant material offers little resistance to your finger, a sticky one requires you to pull hard to free yourself. From data of this kind—a force curve—it is possible to deduce the adhesive and other materials properties of the sample, as long as the data are in the proper form and the tip and cantilever are well calibrated. To begin to understand force curves and thus deepen your understanding of how an AFM tip interacts with the sample, we devote this chapter to some of the necessary background. We must first review the physical concepts of potentials, forces, and concavity. In this chapter's second section, we will examine how force-curve data are displayed and processed, then consider some common problems and their resolutions. Providing the last of the foundations in Chapter 5, the final section reviews the basic mechanical properties of materials.

In **Instrument Lab 5**, you collect dynamic-mode images and force curves.

5.1 Potentials, forces, and concavity

Figure 5.1 displays two common forms of force curves, with photodiode voltage or cantilever deflection on the y-axis and scanner voltage or scanner position on the x-axis. When tip and sample are far away from each other, there is no interaction between them and the force is zero. When they are in firm repulsive contact, at small or negative values of position, the force is positive. Figure 5.1a exhibits no hysteresis, but Fig. 5.1b does, that is, the force during approach of the scanner is different than during the scanner's retraction. The cantilever undergoes rapid changes in deflection, or force, at two different positions, seen by the vertical lines. The physical difference between the figures is that the spring constant of the cantilever is different—all other parameters are the same. Why should there be such a profound change in the acquired data for cantilevers of different spring constants?

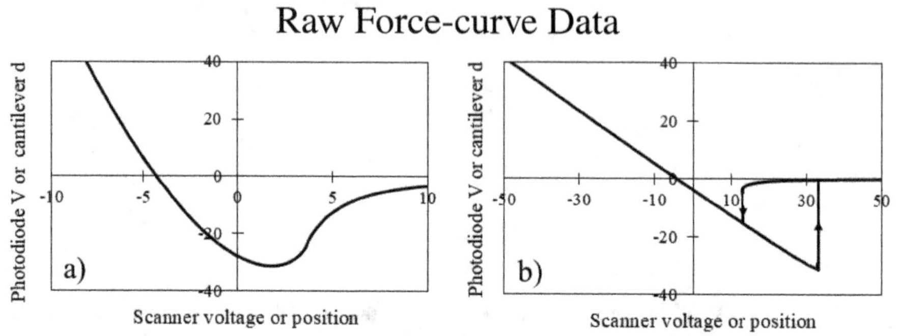

Fig. 5.1 Simulated raw force-curve data taken with (a) a stiff and (b) a weak cantilever. Depending on your AFM, the axes might be photodiode voltage (V) or cantilever deflection (d) and scanner voltage or position. Perplexingly, the tip follows the same path during the approach and retraction of the scanner with the stiff cantilever, but different paths with the weak one, even though the interactions between the tip and sample are the same. Note that the y-axis scales are the same, but the x-axis scales are not.

AFM Lesson 19 at `youtube.com\AtomicForceMicro`

AFM Lesson 19 illustrates characteristic force-curve shapes.

These strange events, called cantilever instabilities, can be understood with the physical ideas of potentials, forces, and concavity, aided by calculus. The most familiar potential is the gravitational potential, in which an object raised in a gravitational field gains gravitational potential energy. When that object is allowed to fall, its potential energy is converted into kinetic energy. The gravitational force on the object is the negative derivative of the potential. The concavity, also called stiffness or the spring constant, of the potential is its positive second derivative or, equivalently, the negative derivative of the force. Let us first look at a simple case of two spring potentials, then examine a more realistic situation that displays some of the same behavior as a weak AFM cantilever during force-curve acquisition.

To progress further we need a well-defined coordinate system, which is shown on the left of Fig. 5.2. The z-axis points up away from the sample surface, towards the AFM's cantilever. Along this axis lie the sample surface at position **s**, the back of the cantilever above the tip at position **c**, and the base of the cantilever at the cantilever holder or head of the microscope, position **h**. The cantilever and the tip-sample interaction can be represented by springs of stiffnesses k_c and k_i, respectively, sketched to the right of the z-axis. The deflection **d** of the cantilever equals **c − h**, and the separation of the tip and sample, or the indentation of the tip into the sample,

δ, equals $\mathbf{c} - \mathbf{s}$. The arrow through the interaction spring and the notation $k_i(\delta)$ are reminders that the spring constant of the interaction is a variable that changes with separation or depth. It is zero when the tip and sample are far away from each other, large and positive when the tip and sample are in firm contact, and negative during the part of the force curve when attractive forces dominate. For the next example, however, you may think of k_i as a constant.

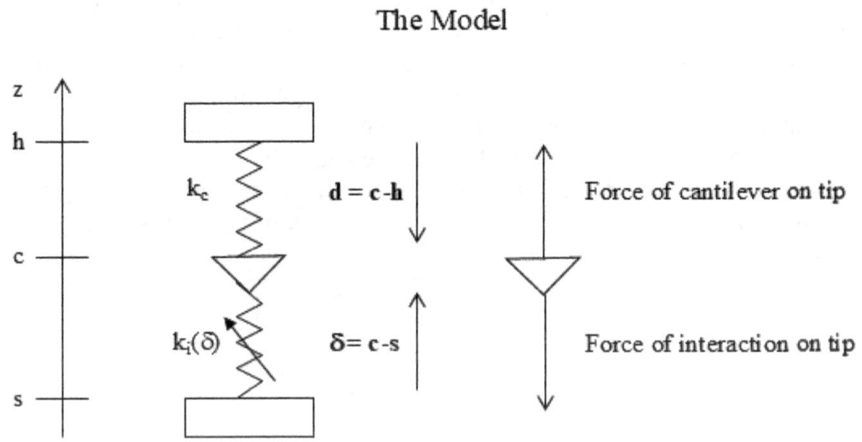

Fig. 5.2 A simple but effective model for interpreting force curves is to represent the cantilever and tip-sample interaction as springs. The cantilever's stiffness (spring constant) is constant. In contrast, the interaction stiffness is variable. The z-coordinate system points vertically up. The vector positions of the cantilever holder, flexible end of the cantilever, and sample surface are denoted by **h**, **c**, and **s**, respectively. The deflection and separation vectors **d** and δ are shown, as well as the forces acting on the tip.

To the right of the spring model are the deflection and depth vectors, **d** and δ, respectively. On the far right is a free-body diagram, showing the forces acting on the tip. In this illustration, the cantilever is bent downwards towards the sample, and the force of the cantilever on the tip is in the positive direction, with the force of the tip-sample interaction on the tip in the negative direction. The directions of the force vectors change when the cantilever is bent upwards due to strong repulsive contact.

5.1.1 An example of two spring potentials

Now, using the coordinate system just described, the bases of two springs of stiffnesses k_c and k_i are placed at the positions **h** and **s**, respectively. They connect at the to-be-determined position **c**, and they represent the cantilever and tip-sample interaction, respectively. The potential energy associated with each one is parabolic in form, centered around the free equilibrium positions **h** and **s**. The total potential

U_{tot} for the tip is the sum of the individual potentials, U_c and U_i.

$$\begin{aligned} U_{tot} &= U_c + U_i \\ &= \frac{1}{2}k_c\,(z-h)^2 + \frac{1}{2}k_i\,(z-s)^2\,. \end{aligned} \tag{5.1}$$

Figure 5.3a shows U_c, U_i, and U_{tot} by the dashed, dotted, and solid lines respectively. You can imagine the equilibrium positions by thinking of how a ball would sit in each potential "valley." A ball in the dotted parabola would lie at s, in the dashed parabola at h, and in the solid parabola at c. All are where the slope of the potential curve is equal to zero. (Note that we have now changed from the bolded vectors to z-components to ease the notation for this one-dimensional case.) Mathematically, to find these equilibria, one takes the negative derivative of the appropriate potential with respect to z and sets it equal to zero. This is the same as saying that for equilibrium, the sum of the forces acting on the ball, or tip, equals zero. The total force is

$$\begin{aligned} F_{tot} &= F_c + F_i = -\frac{dU_{tot}}{dz} \\ &= -k_c\,(z-h) - k_i\,(z-s)\,, \end{aligned} \tag{5.2}$$

and setting the total force equal to zero and solving for z yields the equilibrium position of the free end of the cantilever c,

$$c = \frac{k_c h + k_i s}{k_c + k_i}\,. \tag{5.3}$$

The forces are indicated in Fig. 5.3b, where again the dotted line refers to the sample spring, the dashed line to the cantilever spring, and the solid line to the total force on the tip.

As a reminder, the concavity of the system is the second derivative of the potential, or equivalently, the negative derivative of the force (Eq. 5.4). For two springs, it assumes a very simple form, in which it is always positive ("concave up," as you might have learned in calculus). The concavity is the stiffness or spring constant.

$$\frac{d^2U_{tot}}{dz^2} = -\frac{dF_{tot}}{dz} = k_{tot} = k_c + k_i\,. \tag{5.4}$$

This simple example of two springs has only one stable equilibrium position for c and always has a positive total spring constant, k_{tot}, a situation less common in AFM. The next example is more realistic, inasmuch as multiple equilibria can exist and that the potential is at times "concave down."

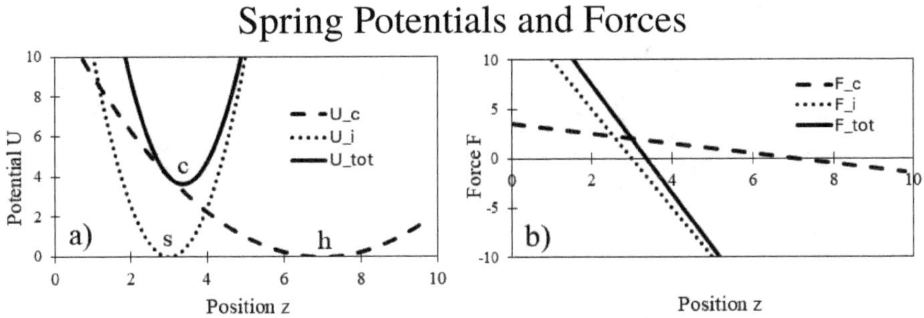

Fig. 5.3 Spring potentials and forces. Plotted in (a) are the spring potentials U_c (dashed line) and U_i (dotted line) and their sum U_{tot} (solid line), all of Eq. 5.1, as a function of position z. Similarly, in (b), the spring forces F_c (dashed line) and F_i (dotted line) and their sum F_{tot} (solid line), all of Eq. 5.2, plotted as a function of position z. In (a), the positions of the sample, flexible end of the cantilever, and cantilever holder are denoted by s, c, and h, respectively, at the minima of the potentials. In (b) the positions correspond to their zero crossings.

5.1.2 A more realistic interaction potential

Building upon the last section, we leave the cantilever potential the same as before, parabolic. We change the tip-sample interaction potential to be of a power-law form, such that its concavity is sometimes negative and sometimes positive. The mathematical form for the total potential U_{tot} is

$$\begin{aligned} U_{tot} &= U_c + U_i \\ &= \frac{1}{2}k_c\,(z-h)^2 + \alpha\left[\left(\frac{\beta}{z-s}\right)^4 - 2\left(\frac{\beta}{z-s}\right)^2\right], \end{aligned} \tag{5.5}$$

where α and β are positive constants.

The potentials U_c, U_i, and U_{tot} are sketched in Fig. 5.4a for a stiff cantilever. The total potential is now negative, zero, or positive, depending on position. As before, one can take the negative derivative with respect to z to find the force, and set it equal to zero to find the equilibrium position c. For the starting position shown for h, at $z = 8$, c is very close to 8. Then the cantilever holder is brought closer to the origin in steps of one z-unit. Shown are the interaction potential and the total potential for each position, labeled with its appropriate h. Throughout the process, there remains one single equilibrium point for the total potential, even though the curve labeled 3 has become a rather distorted parabola. The changes in the total potentials in Fig. 5.4b are a result of a cantilever that is weaker by a factor of five; the interaction potential is the same as in Fig. 5.4a.

In AFM, particularly with a weak cantilever, the system can evolve from having one equilibrium point to having two, then three, then back to two and finally one.

The second derivative of potential with respect to z is variously called the concavity, stiffness, or spring constant, equivalent to the negative derivative of the force. If the value of the stiffness is positive (concave up) at an equilibrium point, then the equilibrium is stable. If the stiffness is negative, (concave down) then the equilibrium

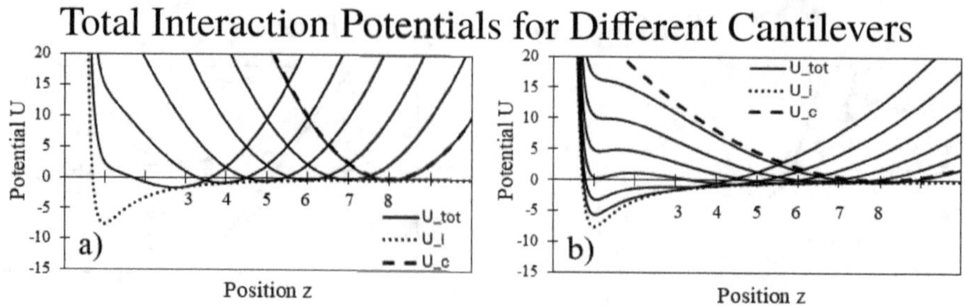

Fig. 5.4 Total interaction potentials U_{tot} (solid lines) of the form of Eq. 5.5 as a function of z for (a) a stiff and (b) a weak cantilever. The numbers 3 to 8 represent different values of h. The tip-sample interaction potential U_i is shown by a dotted line. With the cantilever holder at position $h = 8$, the dashed line is the cantilever potential U_c. For each total potential in (a), there is only one equilibrium position, whereas in (b), there can be more than one.

is unstable. If stiffness equals zero, then the equilibrium is called neutral. Once again, the imagery of a ball rolling on the potential hills and valleys is helpful. A ball in a valley is stable; a ball balanced precariously on top of a hill is unstable. Systems with one stable equilibrium are called monostable, those with two equilibria are bistable.

5.1.3 The origin of cantilever instabilities

During force curve acquisition taken with a cantilever that is stiff relative to the tip-sample interaction, the system evolves such that there is only one stable equilibrium position throughout the entire data acquisition cycle, that is, it remains monostable throughout the force cycle (Fig. 5.4a). The same interaction, studied by means of a weak cantilever (Fig. 5.4b), starts with one stable equilibrium point, then a second neutral one appears, then the neutral one evolves into unstable and stable ones, i.e. the system becomes bistable. Then the system becomes monostable again, all during the approach.

If the stable equilibrium position that the tip occupies disappears, then it is forced to move into the other stable equilibrium. The process is reversed upon tip-sample retraction. The tip will move suddenly from one stable equilibrium point to another when the unstable equilibria disappear. These sudden jumps are called cantilever instabilities. The AFM tip can also move from one stable position to the other if it receives a perturbation—a jolt from someone slamming a door, for example.

A concern about instabilities is that they can make extracting information about the sample properties very difficult, right during a very interesting part of the force curve. Basically, they cause the tip to jump over the important data. On the other hand, good resolution in force is harder to achieve with stiff cantilevers. Topographic imaging could be less accurate because features could be compressed, and worse, you could damage the tip or your sample.

● In **MacroLab 4**, you generate and observe cantilever instabilities. ●

★ In **Computer Lab 4**, you simulate cantilever instabilities. ★

Now that you have developed a sense about the origin of cantilever instabilities, it is time to move on to how force curves are processed, along with some common problems in their interpretation and presentation.

5.2 Force curves

5.2.1 What is a force curve?

A force curve is a plot of the force on the tip from its interaction with the sample as a function of the separation of the tip and sample or the indentation of the tip into the sample. The raw data are obtained when scanning in the x- and y-directions is turned off and the tip and sample are made to first come together, and then apart, by ramping the position of the scan head, so that the raw data are collected as $F_i(h)$. The cantilever first bends downward due to whatever attractive forces act between the tip and the sample. After the tip establishes contact with the sample, the cantilever starts to be pushed upwards, usually deflecting past its free equilibrium point until the operator decides to reverse the z-motion of the scanner. Upon reversal, the tip follows the sample back through the tip's free equilibrium position. Any existing tip-sample adhesion will continue to pull the tip downwards until the force of the cantilever on the tip exceeds the adhesion. Tip and sample break apart, and the tip then returns to its free equilibrium position. Typical forms of force curves as they are acquired can be seen in Fig. 5.1.

In order to glean as much information as possible about the sample, rather than the properties of the cantilever, the data should be in the form of $F_i(\delta)$, the force felt by the tip from its interaction with the sample as a function of separation or penetration depth, instead of $F_i(h)$.

5.2.2 Data processing

What you should appreciate from Sect. 5.1, especially Fig. 5.1, is that your choice of cantilever very much influences the shape of the force curve. In order to be able to say something about the sample, the data should be processed so as to remove as much of the cantilever's influence as possible.

Most instruments collect the data in the format of photodiode voltage as a function of the voltage sent to the scanner, $V_{pd}(V_{scan})$, as in Fig. 5.1. The next step is to present the data in terms of the cantilever deflection as a function of scanner position, $d(s)$ or $d(h)$. (The two dependencies for d represent the scanner carrying the sample or the probe holder, respectively. For ease of discussion, we limit ourselves to the latter situation. Similarly, for ease of discussion, we assume that positive photodiode voltage corresponds to positive force on the tip from its interaction with the sample, i.e., when the cantilever is deflected upwards.) Then deflection is changed to interaction force on the tip as a function of position, $F_i(h)$, by multiplying the deflection d times the spring constant k_c. (For cantilever force F_c on the tip, one multiplies d times $-k_c$. This is the well-known Hooke's Law.) Finally, one can remove some of the influence of the cantilever by processing the curve into a plot of interaction force as a function of separation or indentation depth, $F_i(\delta)$.

5.2.2.1 Converting to force as a function of scanner position

Somewhere within the depths of your instrument's software is a calibration factor that converts the scanner voltage to position. (Section 4.2 describes scanner calibration and typical problems with scanners.) Most commercial AFMs present the raw data already in the form of $V_{pd}(h)$. If not, the user must supply the conversion factor from scanner voltage to position and do the multiplication.

The calibration factor from photodiode voltage to deflection is dependent on both the length of the cantilever and the position of the light spot on the back of the cantilever. If you imagine the scanner moving in the z-direction one nanometer with the tip in firm contact with the sample's surface, the deflection of the cantilever will change by approximately one nanometer, also affecting the angle that the cantilever makes with respect to the light beam. The longer the cantilever, the smaller the angle will be. The photodiode is far more sensitive to angular changes of the cantilever than to changes in its absolute position.

Moreover, as sketched in Fig. 5.5a, the angle that the light spot reflects from the cantilever depends on how far down the cantilever the light spot lies because of how the cantilever's curvature changes along its length. A common method of converting from V_{pd} to d is to take a force curve on smooth, stiff surface (e.g. mica) and assume that the tip-sample interaction is significantly stronger than the cantilever such that the cantilever does all of the compression. Effectively, any change in deflection equals the change in scanner position, such that the scales on the x- and y-axes would now be equal and the slope of the $d(h)$ curve while the tip is in firm contact with the stiff surface equals minus one.

Finally, to convert from deflection of the cantilever to interaction force acting on the tip, the y-axis scale is multiplied by k_c. (For the interaction force acting on the sample, d would be multiplied by $-k_c$.) Frequently, the two steps from $V_{pd}(h)$ to $F_i(h)$ are combined into one, as demonstrated in Fig. 5.5b. The photodiode voltage V_{pd} is plotted as a function of h. The user inputs the cantilever stiffness and measures the slope, $\Delta V_{pd}/\Delta h$, in the firm-contact region of a $V_{pd}(h)$ curve. Then the software performs the conversion

$$F_i(h) = -k_c \left(\frac{\Delta h}{\Delta V_{pd}} \right) V_{pd}, \tag{5.6}$$

where the negative sign accounts for the negative slope.

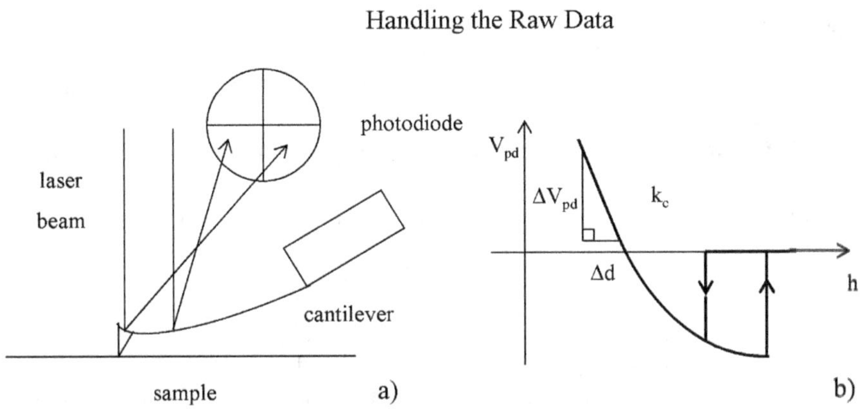

Fig. 5.5 (a) The amount that the reflected light spot moves on the photodiode depends on its position along the cantilever. The closer to the tip, the more the light moves because the cantilever bends more at its tip end than closer to its base. (b) Often raw data are presented as voltage as a function of scanner position, $V_{pd}(h)$. To convert from voltage to force, the cantilever's spring constant must be known and the slope of the firm-contact region of the raw data must be measured. If indeed the tip is in firm contact with the sample, then the change in cantilever deflection Δd is assumed to be the same as the change in scanner height Δh.

If the only property of interest is the tip-sample adhesive force, processing is done. Adhesion, the maximum attractive force or the lowest point of the curve, is the force required to pull the tip off the sample (hence "pull-off force"). If information about the interaction stiffness—and thus the sample's materials properties—is sought, then further data manipulation is necessary. Why? When springs are in series, as in our model of Fig. 5.2, the magnitude of the force on any element in that series is the same, from Newton's Third Law. It also equals the force on a spring that represents the combined spring constants of the cantilever and interaction. This idea is expressed mathematically (considering only magnitudes) as

$$F_i(h) = k_i(c - s) = k_c(c - h) = \frac{k_c k_i}{k_c + k_i}(h - s), \tag{5.7}$$

and if $k_i \gg k_c$, as is often the case, the last term above becomes

$$F_i(h) \rightarrow k_c\,(h - s), \tag{5.8}$$

which conveys no information about the sample at all!

Referring now to Sect. 5.1.3 on cantilever instabilities, the prefactor $k_c k_i/(k_c + k_i)$ on the right-hand side of Eq. 5.7 reveals an alternative way of understanding the variation of the tip-sample interaction force on the tip. When k_i is negative ("concave down" curvature in a potential), the prefactor becomes negative and infinite when $|k_i| = |k_c|$. Upon approach $(h - s) > 0$, and upon retraction $(h - s) < 0$, such that the interaction force is momentarily infinite and negative or infinite and positive.

These two situations account for instabilities in the approach and retraction curves, respectively.

5.2.2.2 Converting to force as a function of separation or indentation depth

To be able to analyze the sample, the data should be in the form

$$F_i(\delta) = -k_i\delta, \tag{5.9}$$

where the negative derivative with respect to δ, is the spring constant of the interface or interaction between the tip and sample, k_i. If the tip is well characterized, sample properties can be measured. Fortunately, we can use our coordinate system to find the separation or indentation depth δ, then replot the $F_i(h)$ graph as $F_i(\delta)$. Referring to Fig. 5.2 and the definitions of d (= $c - h$) and δ (= $c - s$), we see that

$$h - s = (h - c) + (c - s) = -d + \delta. \tag{5.10}$$

With the constant of proportionality between the deflection d and the force $F_i(h)$ equal to k_c, and remembering that the positive direction of d points towards the sample,

$$\delta = h - s - \frac{F_i(h)}{k_c}. \tag{5.11}$$

Now we can replot the $F_i(h)$ data, keeping F_i on the y-axis, and changing the x-axis from h to δ. This has been done to the two data sets of Fig. 5.1, resulting in Fig. 5.6, graphs of the interaction force as a function of separation or indentation depth, $F_i(\delta)$. Now both the x- and y-axes are the same instead of the x-axes being different, as in Fig. 5.1.

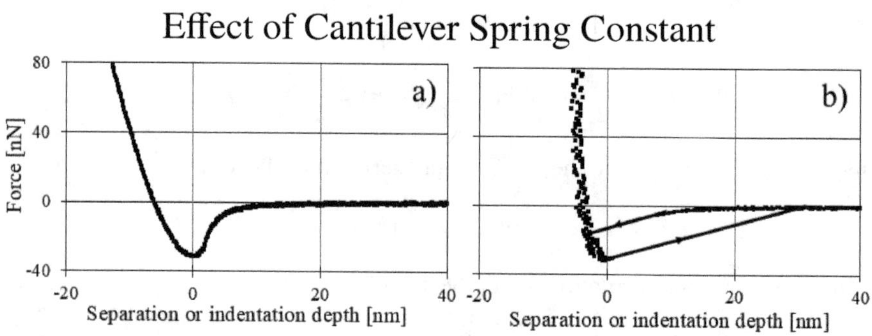

Fig. 5.6 On the left, (a) is a simulated processed force curve, $F_i(\delta)$, the interaction force acting on the tip as a function of separation or indentation depth, taken with a stiff cantilever. In (b) the same interaction is probed by a weak cantilever. Data acquisition with a weak cantilever results in missing data in the force curve, a limited indentation depth, and noise at high force. At high force, h and $F_i(h)/k_c$ become similar in magnitude, and two similar numbers (each with noise) subtracted from each other result in a small noisy signal.

Where does one put the zeroes on the curves? For the y-axis, the answer is clear. Zero interaction force corresponds to zero deflection of the cantilever when it is far from the sample, at its free equilibrium position. The zero point of depth is much less obvious. Should it be during the approach or during the retraction: at the inflection point, at the minimum, at the zero-force position, or somewhere else?

Three reasonable alternatives are at the inflection point of the approach (where contact begins), at the minimum of the approach, or where the surface lies before the forces from the tip might deform it. The inflection point reveals the first stages of contact (Sect. 6.3), but it is more frequently missing due to the use of weak cantilevers. Also, the formation of liquid menisci (Sect. 6.2) can create features that have been mistaken for the beginning of tip-substrate contact. The nominal position of the surface before tip-induced forces might deform it is hard to know before applying contact mechanical models. The practical choice—as a rather rough first approximation—is to put the zero of the x-axis at the minimum of the approach curve. Figure 5.6 shows how processed curves look different depending on whether the data were taken with a stiff or a weak cantilever. The x-axis adjustment is done by using s as an offset. Users with scanners driving the sample position would use h as the offset.

The features of the force curve and their interpretation are best left until a discussion of continuum contact mechanics (Sect. 6.3). From this section on force curves, remember that the negative derivative of an $F_i(\delta)$ curve equals k_i, the stiffness of the tip-sample interaction.

AFM Lesson 20 at `youtube.com\AtomicForceMicro`

AFM Lesson 20 details force-curve processing.

5.2.3 Common problems and questions

Please allow me a short diatribe. Thereafter, I shall discuss five typical problems with and questions about force curves.

5.2.3.1 A short diatribe against weak cantilevers for force curves

Weak cantilevers are useless for measuring materials properties—the ultimate purpose of force curves—except for the pull-off force. "Weak" is relative to the tip-sample interaction, such that a weak cantilever for one experiment might be stiff for another. A "weak" cantilever spring constant might be defined as being one-tenth or less of the magnitude of the maximum interaction spring constant. Physically, this means that most of the compression of the system as the scanner position is changed is accomodated by the weak spring, leaving the tip-sample interface relatively uncompressed. Three problems with weak cantilevers are apparent in Fig. 5.6 and one in Fig. 5.7.

• As mentioned in Sect. 5.1.3, weak cantilevers jump over an interesting part of the force curve, making it difficult, for example, to find the inflection point or the real minimum of the approach.

• During force curve acquisition with a weak cantilever, the scanner moves a lot, the cantilever deflects a lot, and the tip and sample compress a little. To find the indentation depth (tip-sample compression), noisy numbers are subtracted from each other to yield a yet noiser signal. For weak cantilevers, which do not significantly compress the sample, the noise constitutes a higher percentage of the indentation depth than for stiff cantilevers. (Compare Figs. 5.6a and b.)

• The calibration process for a raw force curve, Eq. 5.6, assumes that the interaction is infinitely stiff with respect to the cantilever. If the region over which the calibration was done is included in the processed curve, then the processed curve will have an infinite slope, an unphysical situation. In Fig. 5.6b, where the results for a weak cantilever are shown, the calibration for the curve was done on the net repulsive (positive force) part of the curve. No further apparent indentation takes place as the load (force) on the sample increases. The calibration should be done at loads higher than the loads of interest or on samples that are stiffer than the sample of interest. Extremely weak cantilevers snap into contact with surfaces and give slopes of negative infinity throughout the contact with the sample, clearly a poor depiction of reality.

• Figure 5.7a schematically shows a raw force curve that is also associated with weak cantilevers. The cantilever deflects so much during pull-off that the photodiode becomes saturated and data are lost. Some researchers assume that the retraction curve continues linearly downwards until it reaches the same x-axis value as the instability during retraction (dashed line), then take the pull-off force as the force shown in Fig. 5.7a, a strong assumption. Use a stiff enough cantilever such that the photodiode does not saturate. A sharper tip will also generally result in a lower pull-off force.

AFM Lesson 21 at `youtube.com\AtomicForceMicro`

AFM Lesson 21 describes probe selection for force curves.

● In **MacroLab 5**, you discern how well a weak and stiff cantilever measure the stiffness of an unknown cantilever. ●

★ In **Computer Lab 5**, you see how cantilever stiffness affects the acquired force-curve data. ★

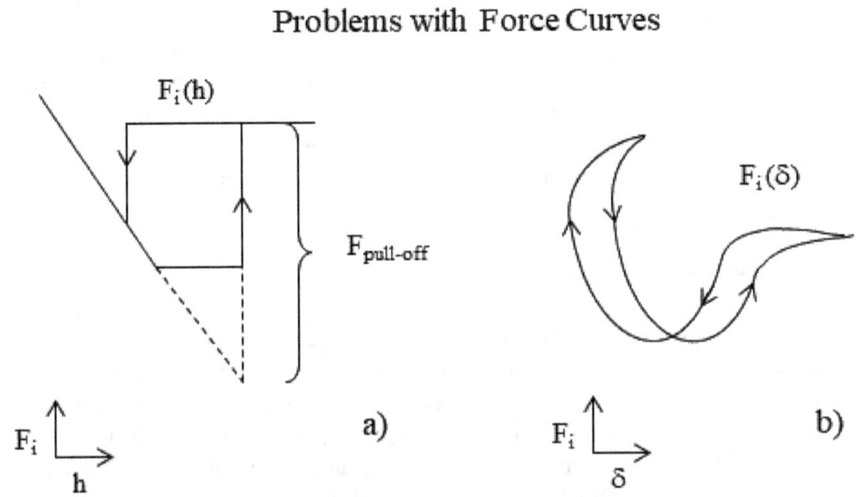

Fig. 5.7 (a) When a cantilever is weak compared to the tip-sample adhesion or pull-off force, the photodiode can become saturated upon retraction, and part of the adhesive region of the curve is lost. (b) There are two problems in this sketch of an $F_i(\delta)$ curve. The first is that the retraction curve is at higher values of indentation depth δ than the approach curve, due to the hysteresis of the scanner. The second is that the top of the force curve has a positive slope because of poor processing.

5.2.3.2 The hysteresis of the scanner

Figure 5.7b displays a sketch of a processed force curve in which the unloading (retraction) part of the curve is at greater values of δ than the loading (approach) part, an unphysical situation. (For this curve to be real, the sample would have to transform into a stiffer sample conveniently just during unloading.) This phenomenon is due to uncompensated scanner hysteresis. One can still use the y-axis force measurements from these bad data sets, but the depth and slope of the curve cannot be trusted. Section 4.2 describes scanners, their hysteresis, and different approaches for correcting the hysteresis.

5.2.3.3 Nonlinear response of the detection system

Occasionally, curvature in the opposite sense to that in Fig. 5.7b is observed in force curves. The reflected light spot at the photodiode is usually not uniform, with the intensity lower around its edges. And depending on how the light is positioned on the cantilever, the reflected intensity can be chopped off at one to three sides. The curvature in the force curve is caused by the reflected light spot moving so far on to one half of the photodiode that the difference in voltage between the two halves is affected by the nonuniform intensity of the reflected light spot. The curvature can be limited by shortening the range of the scanner; large forces can still be obtained by switching to a stiffer cantilever. Any researcher intending to acquire force curves should check force-curve linearity before purchasing an AFM, perhaps by asking an intended supplier to collect and share data from a few typical samples.

5.2.3.4 The misjudged calibration

Another problem in Fig. 5.7b is that during the upper reaches of the curve the slope is positive, when it should be negative or infinite. This artifact arises during the data processing, Eq. 5.6, when it is assumed that the tip-sample contact at the higher loads is significantly much stiffer than the cantilever, $k_i \gg k_c$. Hence in the processed force curve, the slope at the higher loads should be infinite. If the assumption was not completely justified, the curve will tilt slightly to the left (negatively). If the calibration was done inaccurately, the processed curve can flex in the positive direction. Although the calibration should be done on samples much stiffer than the sample of interest or at loads much higher than the load range of interest, these data can be massaged into having an infinite slope at high loads by modifying Eq. 5.11 for δ, where $Error$ is the correction term. It should be of the order of a few percent and can be positive or negative, depending on the origin of the problem.

$$\delta = h - s - \frac{F_i}{k_c}(1 + Error). \tag{5.12}$$

5.2.3.5 A point measurement?

During force-curve acquisition, lateral scanning is disengaged. One might thus be tempted to believe that force curves are point measurements. Unfortunately, a consequence of the small size of typical cantilevers is that they are usually inclined ten to fifteen degrees from the direction that would bring the axis of a tip perpendicularly to the surface. For large movements of the scanner, the tip must slide along the surface approximately in the same direction as the long axis of the cantilever, shown schematically in Fig. 5.8a. The relationship between the sliding distance Δy and the change in height of the scanner Δh is

$$\Delta y \approx \Delta h \tan \Theta, \tag{5.13}$$

where Θ is the angle of repose of the cantilever.

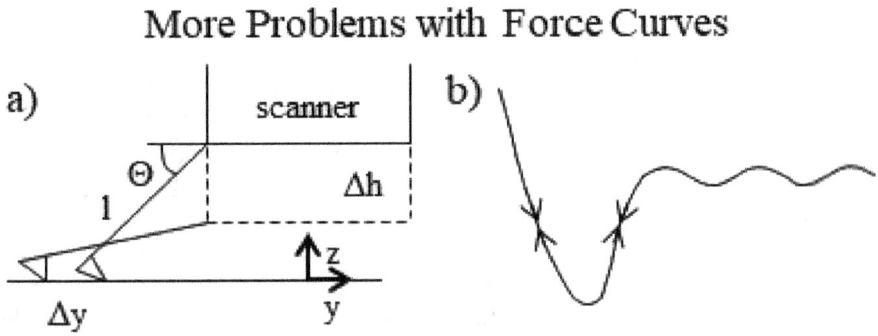

Fig. 5.8 (a) Most cantilevers are inclined ten to fifteen degrees with respect to the sample surface. The result is that a vertical movement of the scanner will result in lateral displacement of the tip. (b) Force curves taken by AFMs equipped with laser detection of the cantilever's position sometimes suffer from interference effects, as seen by the sine waves in the baseline of this sketch.

5.2.3.6 What are the precision and accuracy of force-curve measurements?

Precision is straightforward for the experimentalist to measure. Simply make multiple experiments and calculate the standard deviation. The standard deviation reflects the scatter, or precision, in your experimental data. In comparison, accuracy is the difference between your experimental values and the values obtained by a perfectly calibrated instrument. It is challenging to estimate, and indeed, there are university-level courses devoted to the methodology of estimating accuracy. If your scanner has been calibrated and responds linearly to its input, and if the spring constant of your cantilever has been measured using an accurate approach, then your force data might be good to 10%. If greater accuracy is needed, you must delve into the research literature of how the non-normal inclination of the tip affects the data.

5.2.3.7 Interference

Some AFMs detect the position of the cantilever using a laser. The light from lasers is coherent, meaning that its waves stay in phase as they travel through space. If the light travels two different paths on its way to the photodiode, the amplitudes of the waves add constructively or destructively at the photodiode because of the difference in path lengths traveled by the two beams of light. The resulting amplitude can match the original unsplit beam of light if all of the waves are in phase, or zero if the waves in the two light beams are completely out of phase. This effect is known as interference. In an AFM, if not all of the light is focused on the cantilever, rather, some falls on the sample and is reflected back to the photodiode, then a sinusoidal interference pattern is seen in the baseline of a force curve, as in Fig. 5.8b. This problem can make it difficult to be sure of the zero of force, yet if the wavelength of the laser is known, interference can be used to calibrate the scanner. Please refer to Sect. 4.2.5 for more information.

5.2.3.8 The choice of coordinate system

Sometimes, one sees force curves with the positive and negative directions of the x-axis seemingly reversed, which can be initially confusing. As long as the data are consistent with the author's choice of coordinate system and the definitions of deflection and depth, then the presentation is correct. In this book, the z-axis points upwards normally from the sample surface, d is defined as the difference between the back of the cantilever above the tip minus the position of the base of the cantilever at the cantilever holder or microscope head, $c - h$, and δ is defined as the position of the back of the cantilever minus the sample surface, $c - s$. The plots shown here are consistent with these definitions. Since our scanhead holds the cantilever, it approaches the sample from the positive z direction. A scanner that holds the sample would approach the cantilever from the z direction for the same coordinate system.

AFM Lesson 22 at youtube.com\AtomicForceMicro

AFM Lesson 22 surveys problems with force curves.

This concludes the section on force-curve data processing and problems. In the next section, a summary of mechanical properties—common in materials science classes—provides background for force-curve interpretation.

5.3 Mechanical properties

In this section, we cover the mechanical properties of materials from a traditional viewpoint, which will provide the appropriate background for Sect. 6.3, a simple treatment of contact mechanics for force-curve interpretation. We start by imagining that we will test a macroscopic sample, perhaps of the size and shape of your finger, by grabbing it firmly and pulling on it, as shown schematically in Fig. 5.9a. One might intuitively expect that the sample would first elongate linearly and reversibly, become narrower, then at higher loads undergo nonlinear and nonreversible extension. This is indeed what usually happens, and the process can be monitored by means of a stress-strain curve.

5.3.1 Stress-strain curves

Stress σ is the force F applied to a sample divided by the original cross-sectional area A_0, and strain ϵ is the change in its length Δl divided by its original length l_0. A typical stress-strain curve, as in Fig. 5.9b, starts off linearly at low strain. If the small strain is removed, the sample relaxes back to its original condition.

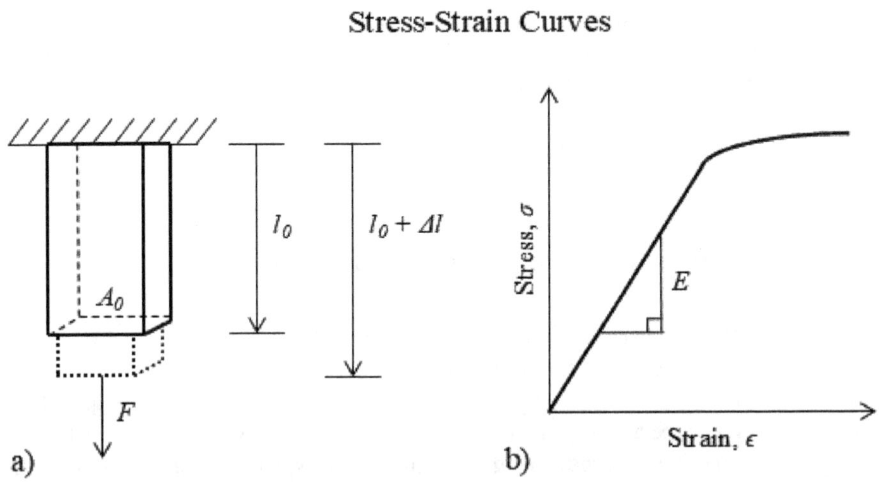

Fig. 5.9 (a) A sample of known original cross-section A_0 is gripped tightly at its ends and pulled by a force F. The sample elongates from its original length l_0 to a new length $l_0 + \Delta l$, narrowing in the process. (b) The stress $\sigma = F/A_0$ plotted as a function of the strain $\epsilon = \Delta l/l_0$. The slope of the stress-strain curve in the initial linear elastic regime is the elastic modulus E.

At higher strains, the stress no longer increases linearly with strain, and additionally, when the strain is removed, the sample remains permanently longer than its original length. The former regime is the elastic regime, the latter, the plastic regime. Some materials return slowly to their original length; this behavior is called anelasticity. Materials that exhibit a mix of elastic and plastic time-dependent behavior are called viscoelastic; for example, many polymers are viscoelastic.

The ratio between stress and strain in the elastic regime of a stress-strain curve is called the Young's or elastic modulus E. The parameter E is the constant of proportionality between stress and strain, much like how the spring constant k relates force and deflection. Stated mathematically, $\sigma = E\epsilon$, with $\sigma = F/A_0$ and $\epsilon = \Delta l/l_0$. The units of both stress and modulus are N/m^2, or Pascals. Strain is unitless.

Just as there is a broad range of materials in the universe, there is a broad range of values for modulus. The elastic modulus of gelatin is in the kilopascal range; soft contact lenses, in the megapascals; polystyrene, in the gigapascals; metals, in the tens or hundreds of gigapascals, diamond, about a terapascal.

5.3.2 Poisson ratio

As the sample lengthens, it narrows, as depicted in Fig. 5.9a. The Poisson ratio υ is the negative of the strain in the direction transverse to the stress divided by the strain in the direction of the stress, that is, $\upsilon = -\epsilon_x/\epsilon_z$. Poisson ratios are always positive;

the negative sign accounts for the decreasing width. In contrast, Δl is positive in the z-direction. Poisson ratios are as small as 0.16 for diamond. Although they are usually around 0.3, they near 0.5 for some polymers. In other words, a diamond sample does not become much thinner while being stretched, but a rubber band does.

5.3.3 Reduced modulus and indentation modulus

Armed with the definitions of elastic modulus and Poisson ratio, we can define how the combined tip and sample would behave if it were one object. In other words, we want to reduce the two-body problem to a one-body problem. You might already have had some experience with the idea of reduction if you have done P. 5.6, in which you found the reduced spring constants of two springs, first in series and then in parallel.

We start by considering two long samples of the same length and cross-sectional area yet of different elastic moduli, bonded together in series by an ideal glue that does not deform. The reduced strain will be the sum of the individual strains of each sample, $\epsilon_r = \epsilon_1 + \epsilon_2$. By means of the relationship between stress and strain, the reduced strain can be expressed as $\epsilon_r = \sigma(1/E_1 + 1/E_2)$. For this case, one would then define the reduced modulus as the inverse of the sum of the inverse moduli, $1/E_r = 1/E_1 + 1/E_2$.

For the situation that we will encounter in Sect. 6.3, in which a tip is pressed into a sample, the reduced modulus takes a more complicated appearance,

$$\frac{1}{E_r} = \frac{1-\upsilon_t^2}{E_t} + \frac{1-\upsilon_s^2}{E_s}, \tag{5.14}$$

where now the subscripts t and s refer to the tip and sample, respectively. The appearance of the Poisson ratios in the numerators is a consequence of the "plane-strain" approximation, in which the strain in the z-direction is assumed small with respect to the strains in the x- and y-directions, as you will investigate in P. 5.14. A full treatment of the three-dimensional interactions between stress and strain begs for help from linear algebra and so will not be endured here.

For simplification of the formulae in Sect. 6.3, Eq. 5.14 is often multiplied by 3/4 to define the indentation modulus K,

$$\frac{1}{K} = \frac{3}{4}\left(\frac{1-\upsilon_t^2}{E_t} + \frac{1-\upsilon_s^2}{E_s}\right). \tag{5.15}$$

5.3.4 Reduced radius

The tip and sample are likely to have different curvatures. Usually, the tip will be small and spherical, and the sample smooth and flat. But we will have more use for our equations in upcoming Sect. 6.3 if they can account for other sample curvatures.

If the tip contacts a small hillock on the sample, you might expect that the tip-sample contact would be less stiff than if the tip were to land on a flat sample. And if the tip were to land in a valley and make a snug fit with the valley walls, the tip-sample contact would be more stiff than a tip and a flat. The hillock would have a positive radius, the flat an infinite radius, and the valley a negative radius. These three situations are shown in Fig. 5.10. Similarly to Eqs. 5.14 and 5.15 above, the reduced radius R_r is the inverse of the sum of the inverses of the radii.

$$\frac{1}{R_r} = \frac{1}{R_t} + \frac{1}{R_s}. \tag{5.16}$$

This equation might remind you of the equivalent, or reduced, resistance of two resistors in parallel. And indeed, they are derived in very similar ways, as you will discover in P. 5.18. For the resistors, one sums the currents in each of the two resistors, then expresses both the reduced current and individual currents in terms of the resistances and the voltage drop. Since the voltage drop is everywhere the same, it can be factored out, leaving the definition of the reduced resistance.

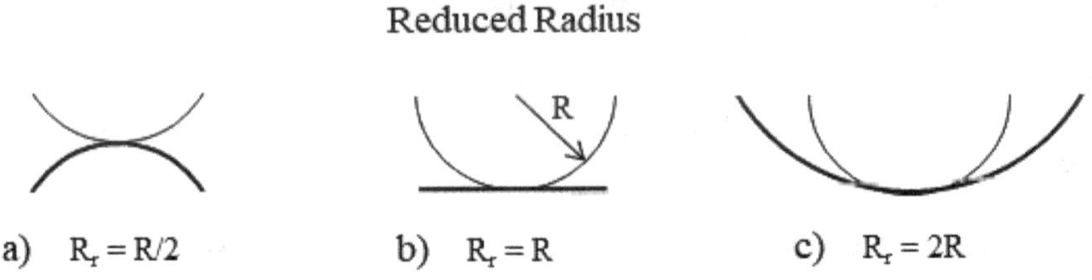

Fig. 5.10 The reduced radius R_r for a tip of radius R (thin lines) and a sample (thick lines) of three different curvatures: (a) R, (b) ∞, and (c) $-2R$.

5.3.5 Contact radius

Let us imagine that a sphere is indented into a flat surface to a depth δ, as shown in Fig. 5.11a. The sphere and flat remain in contact with each other until a depth $\delta/2$, which corresponds to a point on the sphere that is a radius a out from its vertical axis. Assuming that the depth is small in comparison to the radius, the relationship between a and δ can be approximated by (see P. 5.17)

$$a^2 \approx R\delta. \tag{5.17}$$

Be aware that tips of different geometries have different relationships between contact radius and depth.

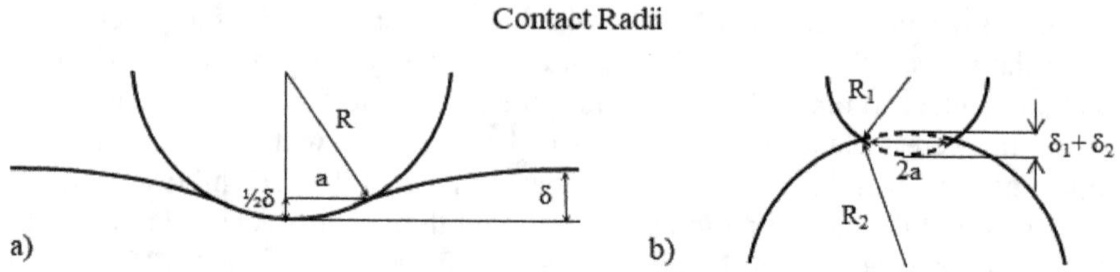

Fig. 5.11 (a) A spherical tip of radius R is indented into a compliant sample to a depth δ. The radius of contact between the two bodies is a. They loose contact at a depth of $\delta/2$. (b) Two compliant spheres of different radii form a mutual circular contact of radius a, as described in P. 5.16. The total deformation is $\delta_1 + \delta_2$.

5.3.6 Work of adhesion

Work of adhesion is the magnitude of the energy required, per unit area, to separate two smooth flat surfaces from contact to infinity. Refer back to the tip-sample interaction shown in Fig. 5.4 by the dotted lines. The change in a variable is its final minus its initial values. Therefore the change in potential by pulling the tip and sample apart from their potential minimum (zero force) to infinity is $\Delta U = U_\infty - U_{\min} = 0 - (-\alpha) = \alpha$, from the definition of U_i in Eq. 5.5, in which the prefactor α is the depth of the potential well. To obtain the work of adhesion, α would be divided by the contact area.

The symbol we use for work of adhesion is ϖ, and its units are J/m^2, or equivalently, N/m. Again, there is a wide range of behavior in this world, from the very low work of adhesion of carbon tetrafluoride at approximately 40 mJ/m^2, to clean metals at several J/m^2. In fact, clean metals in vacuum will automatically bond to each other. The work of adhesion can be related to the strength of van der Waals interactions, which we will explore in Sect. 6.2.1.

AFM Lesson 23 at `youtube.com\AtomicForceMicro`

AFM Lesson 23 supplies this background for force-curve interpretation.

Now with with the basics of force curves and materials properties established, in the next chapter we shall move on to why force curves have both attractive and repulsive regions, as well as how to model them.

5.4 Learning objectives, further reading, and problems

! Learning objectives for Chapter 5

From the material in Chapter 5, the reader should now be able to:

5.1 Know the basic physical principles behind force curves—the relationships among potentials, forces, and concavity.
5.2 Be able to determine if an equilibrium is stable, unstable, or neutral, both mathematically and by inspecting a potential diagram.
5.3 Define monostability and bistability.
5.4 Explain the origin of cantilever instabilities.
5.5 Define a force curve.
5.6 Understand how raw force-curve data are converted to processed data.
5.7 Discuss the limitations of weak cantilevers for force-curve aquisition.
5.8 List and describe common problems with force curves.
5.9 Compare and contrast elasticity, plasticity, and viscoelasticity.
5.10 Be able to sketch and discuss a stress-strain curve; define σ, ϵ, and E.
5.11 Define υ, E_r, and K.
5.12 Define R_r, a, and ϖ.

Further Reading

✓**History.** Burnham, N. A., Colton, R. J., & Pollock, H. M. (1991). Interpretation issues in force microscopy. Journal of Vacuum Science & Technology A: Vacuum, Surfaces, and Films, 9(4), 2548–2556.

✓**History.** Sugisaki, K., Nakagiri, N., & Kinjo, Y. (1999). Artifacts in adhesion force images obtained by force curve mapping. Langmuir, 15(15), 5093–5097.

✓**Book.** Callister Jr, W. D., & Rethwisch, D. G. (2020). Materials science and engineering: an introduction. John Wiley & Sons.

✓**Book.** Halliday, D., Resnick, R., & Walker, J. (2023). Principles of physics. John Wiley & Sons.

? Problems for Chapter 5

5.1 (a) How many equilibria are there for each of the total potentials in Figs. 5.4a and b? (b) Is there a neutral equilibrium? If so, where? (c) Which curves would you call monostable, and which ones would you call bistable?

5.2 In Fig. 5.4b, at what values of h must cantilever instabilities occur for approach and retraction?

5.3 Derive the force and spring constant from the spring potential, $U_s(x) = \frac{1}{2}kx^2$.

5.4 For the function $f(x) = x^2 + x^3$, find the equilibria and demonstrate mathematically whether they are stable or unstable. Sketch the function and mark the equilibrium points.

5.5 (a) Assuming $s = 0$ for simplicity, derive the force and (b) spring constant from U_{tot}, Eq. 5.5.

5.6 (a) Referring to Fig. 5.2, $|d| + |\delta| = |h - s|$. Since the magnitude of the force in Fig. 5.2 is everywhere the same by Newton's Third Law, substitute F/k_c for $|d|$, F/k_i for $|\delta|$, and F/k_r for $|h - s|$. Solve for k_r, the reduced spring constant for the sample-tip-cantilever system, which appears in Eq. 5.7. (b) As a comparison, now imagine that instead of k_c and k_i arranged in series, as in Fig. 5.2, they are aligned in parallel. They will be compressed the same amount, d, and the force on each spring will be different. The total force will be the sum of the individual forces on the springs. Solve for the reduced spring constant $k_r^{||}$ in this case.

5.7 In the limit of $k_i \gg k_c$, show how Eq. 5.7 becomes Eq. 5.8.

5.8 Figure 5.2 shows the model used to interpret force curves when the cantilever is bent down towards the sample due to attractive forces. Make an equivalent sketch for the case of repulsive forces, showing the z-axis, the positions of h, c, and s, the springs k_c and $k_i(\delta)$, the vectors for deflection and indentation, and the directions of the forces of the cantilever and sample on the tip.

5.9 Consider Eq. 5.11 and allow the position of the sample surface s to be zero. (a) If a small change in the cantilever holder Δh exactly equals a small change in deflection Δd, what is the change in separation or indentation depth $\Delta\delta$? What consequence does this have for either a $d(\delta)$ or an $F_i(\delta)$ curve? (b) If instead there has been a small error, and $\Delta h = \Delta d(1 + Error)$, what is $\Delta\delta$? What is the consequence for a $d(\delta)$ or an $F_i(\delta)$ curve? (c) If at an instability in an $F_i(h)$ curve we observe no measurable change in Δh while Δd changes significantly, what is the change in $\Delta\delta$? What is the consequence for a $d(\delta)$ or an $F_i(\delta)$ curve?

5.10 Consider the effect of friction on a raw force curve with the tip in firm contact with a stiff sample, in the form of $d(h)$. The slope of the curve as the tip slides along the surface is expressed by

$$Slope = -\frac{1 - \frac{2\eta}{l}\left(\frac{\tan\theta \pm \mu}{1 \mp \mu \tan\theta}\right)}{1 - \frac{3\eta}{2l}\left(\frac{\tan\theta \pm \mu}{1 \mp \mu \tan\theta}\right)},$$

where η is the tip height, l is the cantilever's length, θ is the cantilever's angle of repose (assumed small), and μ is the coefficient of friction (assumed small).

The upper symbol in ± or ∓ refers to loading, the lower one to unloading. (a) Employ the binomial theorem to simplify the expression for slope to an approximate expression that is first-order in η, l, $\tan\theta$, and μ. (b) Which has the steeper slope, loading or unloading?

5.11 Starting from Fig. 5.8a, derive Eq. 5.13, the sliding distance of the cantilever. Assume that the cantilever does not bow, i.e., the effective length of the cantilever, l, does not change during sliding. It is the hypotenuse of a triangle formed with sides h and y at the beginning of a force curve, and at the closest approach of the scanner, $h - \Delta h$ and $y + \Delta y$.

5.12 Imagine a sample of dimensions x_0, y_0, and z_0. The sample is pulled along the z-axis. The strains remain small. (a) Express the new volume in terms of its original dimensions and its strains ϵ_x, ϵ_y, and ϵ_z . You may drop small terms. (b) Now find the volumetric strain, that is, the change in volume divided by the original volume. (c) Rewrite your answer, expressing the lateral strains in terms of the Poisson ratio υ and the longitudinal strain ϵ_z. (d) What is the highest value of υ that makes sense? A rubber band might have a Poisson ratio approaching this value.

5.13 (a) What is the reduced modulus E_r of Eq. 5.14 for a tip that is significantly stiffer than the sample? (b) What is E_r if the tip and sample have the same moduli E and Poisson ratios υ? (c) And finally, what is E_r if the sample is significantly stiffer than the tip?

5.14 (a) Find the approximate strain underneath the tip of Fig. 5.11a. The original length of the sample surface beneath the tip is a. By using the Pythagorean and Binomial Theorems on a triangle of sides a and $\delta/2$, with $a \gg \delta/2$, find how much the sample surface elongates, and solve for the strain that is mostly within the plane, ϵ_x. Use Eq. 5.17 to express your answer in terms of R. (b) If the sample is 800 μm thick, δ is 8 nm, and R is 1 μm, is the "plane-strain" approximation that leads to Eq. 5.14 valid? (c) What if, instead, the sample is 800 nm thick?

5.15 The maximum strain beneath a spherical AFM tip is at the perimeter of the contact. For Hertzian geometry, the perimeter is at depth $\delta/2$; see Fig. 5.11a. If the sample surface at the perimeter is stretched a distance ds along the surface of the tip, what is the strain $\epsilon_{max} = (ds - dx)/dx$? *Hints:* Use geometry to determine how the angle of the surface at the perimeter is related to $\theta = \arcsin(a/R)$. Then use $ds^2 = dx^2 + dz^2$.

5.16 Using Eq. 5.16, confirm that the values for the reduced radius R_r given in Fig. 5.10 are correct.

5.17 Imagine a sphere of radius R pressed into a smooth, flat, compliant surface, as in Fig. 5.11a. The total deformation of the sample is δ, and the sphere and flat loose contact with each other at a depth of $\delta/2$. The deformation is much less than the tip's radius, $\delta \ll R$. By use of the Pythagorean Theorem, find the square of the contact radius a, Eq. 5.17. You may drop small terms.

5.18 Two compliant spheres of different radii R_1 and R_2 are pushed together to form a mutual circular contact of radius a, as shown in Fig. 5.11b. Derive the expression for the reduced radius R_r, Eq. 5.16, starting with $\delta_1 + \delta_2 = \delta_{tot}$, then invoking 5.17.

5.19 Look online for two publications that use force curves, one with the curves displayed as a function of scanner position and one with the curves displayed as a function of separation or indentation depth. Send the links to your instructor. In a few sentences, be prepared to tell the class what the authors of the papers wanted to learn.

5.20 Lab question. Assuming that you have acquired a force curve with a weak cantilever, label these parts of it: baseline, instability upon approach, instability upon retraction, firm contact region, force of adhesion. If the maximum photodiode voltages were not exceeded, sketch how they would look if they were.

5.21 *Challenge Problem.* In P. 5.6, we found the expressions for k_r and $k_r^{||}$ for series and parallel springs, respectively. Yet in our simple example of Sect. 5.1.1, the response (Eq. 5.4) is that of parallel springs, even though the springs are configured in series. Explain why in words.

5.22 *Challenge Problem.* Set $s = 0$ in the the power-law potential $U_i(z)$ in Eq. 5.5. (a) Starting with $U_i(z)$, what are the expressions for $F_i(z)$ and $k_i(z)$? (b) Find the position of the equilibrium, z_0. (c) Calculate $U_i(\beta)$ and $k_i(\beta)$. (d) At what value of z does $k_i(z) = 0$? This is the inflection point of the potential. (e) Determine $U_i(\sqrt{5/3}\beta)$ and $F_i(\sqrt{5/3}\beta)$. (f) Sketch $U_i(z)$ and $F_i(z)$, indicating the potential, force, and z-values of the equilibrium and inflection points. (g) What are the meanings of α and β?

Chapter 6
Tip-sample interactions

Abstract AFMs were developed for imaging. Force-curve measurements are a secondary use. AFMs' two principal limitations for force-cruve acquisition are that the cantilever is not perpendicular to the vertical motion of the scanner (Sect. 5.2.3.5) and that silicon and silicon nitride tips (Sect. 4.1.1) are not robust. Nonetheless, there is value in having a fast, high-resolution, three-dimensional imaging tool that can also make passable force-curve measurements that reveal the mechanical properties of the sample. And AFMs are relatively inexpensive and have thus become fairly common. Hence, we focus on using AFM to collect such data. The reader should be familiar with the contents of Chap. 5 before delving into an introductory section on how AFM can probe the first stages of contact. This short section is followed by three more substantial ones on surface forces and adhesion, continuum contact mechanics, and molecular dynamics, all of which are helpful to the interpretation of force-curve data and thus serve to deepen understanding of how an AFM works.

6.1 The first stages of contact

The use of AFM for force-curve acquisition starting in the late 1980's opened an era of the study of the very first stages of contact. Previous approaches had explored attractive and adhesive interactions (surface forces apparatus) and fully repulsive contact (hardness tester), yet it awaited the AFM's combination of high force resolution and sharp tips for the systematic study of a regime where both attractive and repulsive forces are important. Figure 6.1 displays a sketch of the regimes where AFM and related instruments are most appropriate. The surface forces apparatus and the hardness tester ("indenter") will not be covered here, although the physics and materials science developed for them are foundational to Sects. 5.3, 6.2, and 6.3.

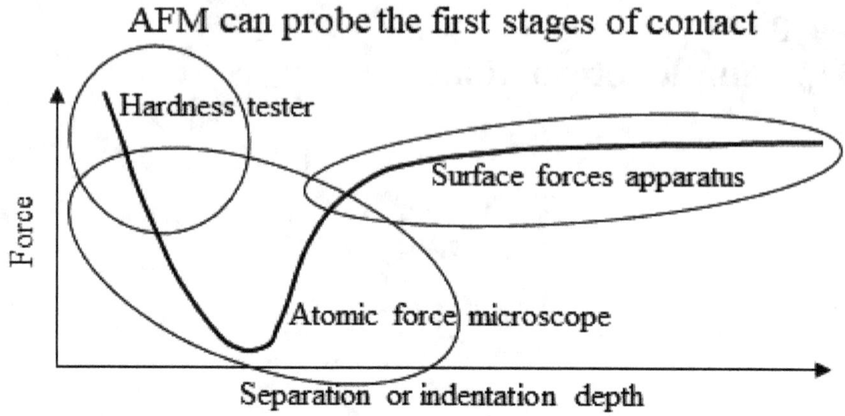

Fig. 6.1 Sketch of a typical force curve, force as a function of separation or indentation depth. Earlier instruments, notably the hardness tester and the surface forces apparatus, had probed fully repulsive contact or attractive and adhesive forces, respectively. The combination of AFM's high force and lateral resolution allows the regime of the first stages of contact to be explored.

> **AFM Lesson 24 at `youtube.com\AtomicForceMicro`**
>
> AFM Lesson 24 covers surface forces.

6.2 Surface forces and adhesion

There are many reasons that the cantilever bends forward during the attractive part of the force curve. There could be long-ranged forces between the tip and sample due to electrostatic, electrodynamic, or magnetic interactions. These forces can change radically if liquid layers are present or if the tip-sample system is immersed in a fluid. More than one force can operate at once. Forces before contact can even be repulsive, which is why the term "surface forces" is more general than "attractive forces."

With all of the possibilities, it can be difficult to discern which force or forces dominate before the tip and sample make contact. The clues to the operative forces are the curve's sign, magnitude, and functional dependence on separation. A typical magnitude for attractive, precontact forces lies between 1 and 100 nN.

The word adhesion is used after the tip and sample have made contact. Generally, adhesive forces are greater in magnitude than precontact attractive forces. During contact, chemical bonding adds to any pre-existing attraction, and plastic deformation can change the shape of the curve. If no chemical bonding and no plastic deformation occur, the approach and retraction traces of a force curve overlie each other.

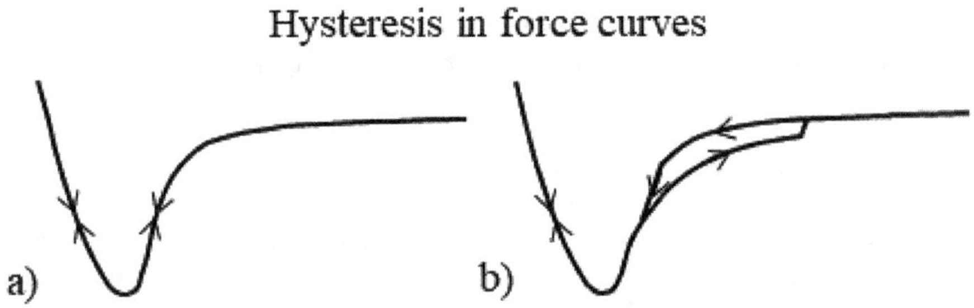

Fig. 6.2 (a) Sketch of a force curve, $F(\delta)$, for a van der Waals interaction, which is not intrinsically hysteretic. In the absence of plastic deformation or bonding during contact, the approach and retraction traces should be the same. (b) $F(\delta)$ for a meniscus-capillary interaction, which *is* intrinsically hysteretic. The formation of a meniscus draws the tip suddenly towards the sample, and the formation of a capillary upon retraction causes the force to remain more negative than during the approach, until the "liquid bridge" breaks.

Entire texts and entire semester-long courses have been devoted to the subject of surface forces and adhesion. Here, we pick two common surface forces that you will probably encounter. They are van der Waals interactions, and meniscus and capillary formation of surface liquid films. If only van der Waals interactions are responsible for the precontact bending of a stiff cantilever, and no chemical bonding or plastic deformation occurs during contact, the approach and retraction parts of the force curve will be the same, as shown in Fig. 6.2a. In contrast, the formation of meniscii and capillaries is inherently hysteretic, that is, the approach and retraction traces are different, as seen by the loop in Fig. 6.2b. This section concludes by considering how a weak cantilever can prevent the determination of which forces are operative and by making the connection between force curves and the discussion in Sect. 5.1 on potentials, forces, and stiffnesses.

6.2.1 Attraction by van der Waals interactions

Back in Sect. 5.1, we introduced a power-law potential, Eq. 5.5, that was a more realistic, yet pedagogical, representation of the potential between a tip and a sample than the spring potential. Compare Eq. 5.5 with this potential,

$$U_{LJ} = U_0 \left[\left(\frac{r_0}{r} \right)^{12} - 2 \left(\frac{r_0}{r} \right)^{6} \right], \tag{6.1}$$

which is known as the Lennard-Jones, or "6–12", potential for the interaction between two atoms. The parameter r is the interatomic distance, and U_0 and r_0 are positive constants. This potential, although quite simple, has been used extensively in molecular modeling and, in some cases, provides a remarkably good fit to experimental data.

In this book, we discuss the second term, in which the attractive part of the potential is proportional to the inverse sixth power of the separation between two atoms.

6.2.1.1 Several physical systems have $1/r^6$ potentials

Part of the reason that the $1/r^6$ potential has acquired such importance in that it accounts for several different physical phenomena by a single, simple functional form. Both permanent dipoles and polarizable atoms and molecules are involved. A dipole is a structure that is negatively charged on one end and positively charged on the other, while its net charge remains zero. A water molecule is a good example of a permanent dipole, in which the oxygen atom pulls the electrons away from the hydrogen atoms, creating negative and positive sides of the water molecule.

Polarizability describes how easily an atom or molecule can be induced to become a dipole. In an external electric field, the average position of the electrons in an atom can shift from being centered around the nucleus to being centered slightly in the direction opposing the direction of the electric field. (Remember that the electric field points in the direction a positive charge would move.)

The $1/r^6$ potential applies to: (i) freely rotating permanent dipoles interacting with each other, for example, water molecules, (ii) rotating or fixed permanent dipoles interacting with polarizable atoms or molecules, and (iii) polarizable atoms or molecules that become polarized in concert with their neighbors, due to rapid and collective fluctuations in the average position of electrons around the nuclei. Since all atoms and molecules are polarizable to lesser or greater extent, the $1/r^6$ potential applies to all substances.

The Coulomb force laws that you have encountered in your education so far have an inverse-squared dependence on distance. Why would dipoles and induced dipoles behave so differently?

6.2.1.2 Why do two dipoles attract as the inverse sixth power of their separation?

An exact treatment of the origins of the van der Waals potential requires a thorough understanding of quantum mechanics and the optical properties of materials. We will step through a simpler treatment that, while inexact, should give you an understanding of where the $1/r^6$ potential comes from.

Imagine a dipole of charge q_1 and length a_1 aligned along the z-axis, centered about the origin. It creates an electric field $\mathbf{E}$ at an atom position $\mathbf{r}$, as sketched in Fig. 6.3a. That polarizable atom itself becomes a dipole, with the positive charge being forced to negative z-values, creating a nonzero force acting on the original dipole, Fig. 6.3b. Since the charges of same sign are slightly farther away from each other than the charges of the opposite sign, there is a small net force acting on the original dipole to the right. By Newton's Third Law, the same magnitude of force acts to the left on the second dipole; those force vectors are not shown. Once we obtain the force between the dipoles, it is a simple matter to solve for the potential.

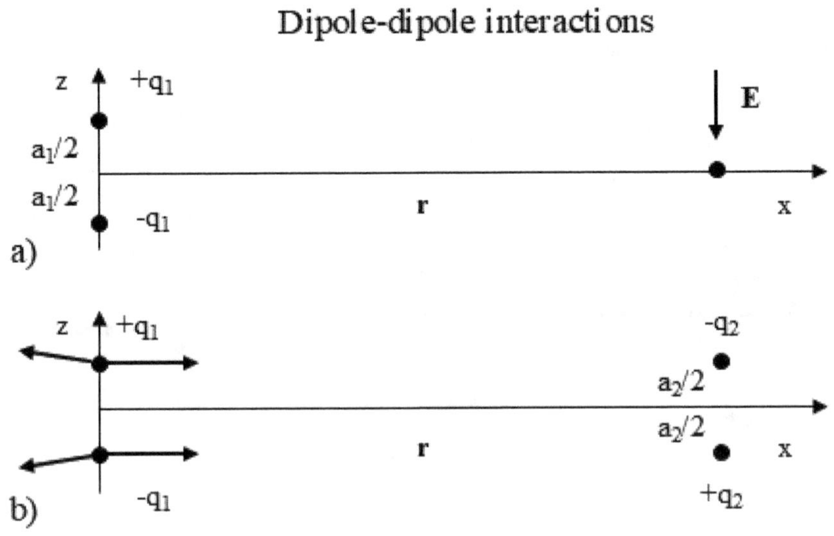

Fig. 6.3 (a) A permanent dipole is aligned along the z-axis, centered on the origin, with the two charges a distance a_1 away from each other. It creates an electric field, proportional to $1/r^3$, at an atom positioned at **r** along the x-axis. (b) The electric field induces a dipole in the atom, which is also proportional to $1/r^3$. The two charges of the original dipole at the origin are subjected to forces between them and the charges of the induced dipole, signified by the four arrows. By Newton's Third Law, these same forces act on the induced dipole at position **r** (force vectors not shown). The text explains why the potential for the attractive force between the dipoles is proportional to $1/r^6$.

Now here come the proportionalities. (You will derive the exact expressions in P. 6.30.) The electric field at position **r** is proportional to $1/r^3$. Because the polarizability of an atom is directly proportional to the electric field, the magnitude of the dipole q_2a_2 at **r** is also proportional to $1/r^3$. The magnitudes of the forces between the pairs of charges acting on the dipole at the origin are similar, but not exactly the same. We want to add the four forces from the dipole at position **r** acting on the dipole at the origin. Since the distance between dipoles is much greater than the separation a_i between the charges within the dipole, the binomial theorem allows us to expand the expression for the total force in terms of a_i/r, where for simplicity, we assume that $a_1 = a_2$. By doing so, the force goes as q_2a_2/r^4, and because of the dependence of q_2a_2 on $1/r^3$, the force in fact goes as $1/r^7$. Integrating and negating for the potential, one obtains $U_{a-a} = -B/r^6$, where the constant B is for you to determine in P. 6.30.

Now we have established the inverse-sixth power law for the potential between a fixed permanent dipole and an induced one. A rotating dipole with an induced one, two induced dipoles, and permanent dipoles that are free to rotate have the same functional dependence on r, although their prefactors differ.

6.2.1.3 Why do different shapes attract with different functional dependencies?

The attractive interatomic van der Waals potential, $U_{a-a} = -B/r^6$, applies to atoms and molecules, quite a few size scales below the typical sizes of tips and samples in AFM. A reasonable approximation for interactions between micro- or macroscopic bodies is to sum up over all the dipoles or inducible dipoles in the bodies, for example, a hemispherical tip and a large enough sample such that its dimensions dwarf that of the tip. We are about to embark on a series of volume integrations...and you thought you would never have any use for calculus!

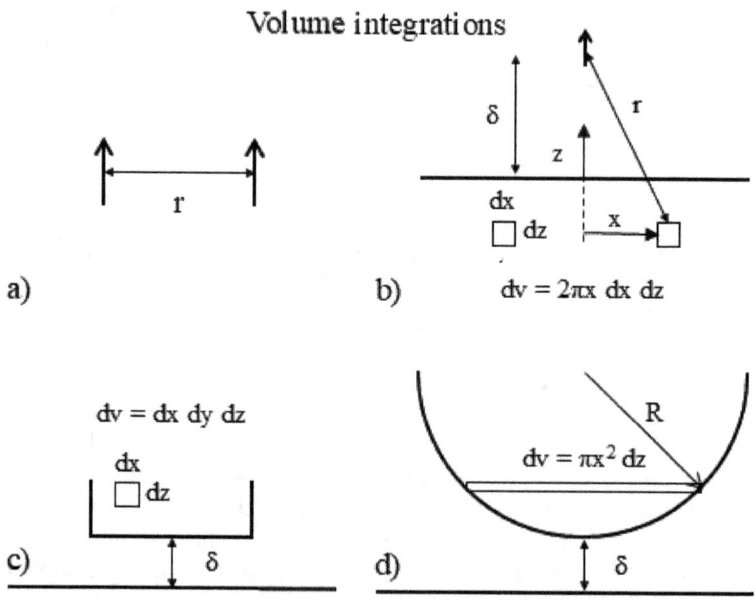

Fig. 6.4 (a) The van der Waals potential between two atoms is proportional to $1/r^6$. (b) A volume integration over an infinite half-space for a lone dipole a distance δ away yields a potential that is proportional to $1/\delta^3$. (c) Another volume integration over a flat of limited area A results in a potential per unit area that is proportional to $1/\delta^2$. (d) A volume integration over a hemisphere results in a potential that is proportional to $1/\delta$.

The first step is to integrate the inverse-sixth power potential between a single dipole and an "infinite" half-space of dipoles situated a distance δ away. Figure 6.4a shows two dipoles, and Fig. 6.4b displays the configuration for the integration. An annular differential volume, $dv = 2\pi x\, dx\, dz$, multiplied by the number of dipoles per unit volume, ϱ_s, is convenient for this purpose. Within the annulus, all of the dipoles are a distance r away from the lone dipole. The integration (P. 6.2) yields the atom-sample potential $U_{a-s} = -\pi\varrho_s B/6\delta^3$.

Before proceeding to the more difficult sphere-sample integration, it is good practice to do the easier integration for a small flat of area A positioned a distance δ away from the infinitely large sample, as displayed in Fig. 6.4c. The integrations continue with the number of dipoles in the differential volume equal to $\varrho_f A\,dz$. The potential per unit area between the flat and sample is $U_{f-s}/A = -\pi\varrho_f\varrho_s B/12\delta^2$. By defining a new constant, the Hamaker constant, to be $H = \pi^2\varrho_f\varrho_s B$, the flat-sample potential per unit area becomes $U_{f-s}/A = -H/12\pi\delta^2$. A typical value for H is 10^{-19} J. Because work and potential are the opposites of each other, the work of adhesion per unit area as defined in Sect. 5.3.6 is thus $\varpi = H/12\pi\delta_0^2$, where δ_0 is the separation at contact.

And now, our model for a typical AFM tip-sample system—Fig. 6.4d—a tip with a hemispherical end large enough such that at the upper part of the hemisphere where it joins the shaft of the tip, the van der Waals interactions are small. The infinite sample is flat and smooth. The tip radius is big compared with the separation, $R \gg \delta$. The Pythagorean Theorem sets the relationship between the square of the distance out from the central axis of the tip to be $x^2 \approx 2R(z-\delta)$, in a similar way as to your solution to P. 5.17. The number density of dipoles in the tip, ϱ_t, times the differential volume for the sphere, $\pi x^2\,dz$, is thus $2\pi\varrho_t R(z-\delta)\,dz$. Starting from the atom-sample potential and integrating from δ to ∞, the tip-sample potential is found to be $U_{t-s} = -HR/(6\delta)$.

If at $\delta + R$, the van der Waals interaction is not negligible compared to the strength of the interaction at δ, the integration is done from δ to $\delta + R$, and a second integration must be done to include the geometry of the shaft of the tip.

Problem 6.2 will ask you to show all the steps of these integrations for the potentials. Problem 6.3 asks you to find the van der Waals force between a typical AFM tip and a sample, a useful expression for estimating the tip-sample force in many situations.

6.2.1.4 A treatment based on quantum mechanical and optical properties

Now we will give you a taste of a higher-level treatment. Below are equations for the Hamaker constant, calculated in a more sophisticated way than that above, based on fully fledged quantum mechanics and the optical properties of materials. Even though the equation for the Hamaker constant is different after proper treatment, the distance dependence for the various geometries discussed in the previous paragraphs stays the same.

The Hamaker constant becomes the sum of two terms. The first term, $H_{\omega=0}$, represents the contribution of permanent dipoles. No fluctuations of electrons are involved, so it called the zero frequency or static term. The second term, $H_{\omega>0}$, represents the contribution of fluctuating dipoles, i.e. the instantaneous movements of electrons around their nuclei, and as such it is called the high frequency or dynamic term. Depending on the materials involved, the dynamic term can dominate over the static one or they can be of similar magnitudes (P. 6.4).

In the equations below, k_B is Boltzmann's constant, T is temperature in Kelvin, the ε are dielectric constants, $\hbar$ is Planck's constant divided by 2π, ω_e is the excitation frequency of the electrons, the n are indices of refraction, and the subscripts t, s, and m, refer to the tip, sample, and intervening medium, respectively.

$$H = H_{\omega=0} + H_{\omega>0}, \tag{6.2}$$

$$H_{\omega=0} = \frac{3}{4}k_B T\left(\frac{\varepsilon_t - \varepsilon_m}{\varepsilon_t + \varepsilon_m}\right)\left(\frac{\varepsilon_s - \varepsilon_m}{\varepsilon_s + \varepsilon_m}\right),$$

$$H_{\omega>0} \approx \frac{3\hbar\omega_e}{8\sqrt{2}} \frac{\left(n_t^2 - n_m^2\right)\left(n_s^2 - n_m^2\right)}{\left(n_t^2 + n_m^2\right)^{\frac{1}{2}}\left(n_s^2 + n_m^2\right)^{\frac{1}{2}}\left[\left(n_t^2 + n_m^2\right)^{\frac{1}{2}} + \left(n_s^2 + n_m^2\right)^{\frac{1}{2}}\right]}.$$

It is instructive to consider special cases of the above to see if H is positive, resulting in an attractive van der Waals force, negative, resulting in a repulsive force, or zero. You will do so in P. 6.5.

AFM Lesson 25 at youtube.com\AtomicForceMicro

AFM Lesson 25 details van der Waals forces.

6.2.2 Meniscus and capillary formation

Another common tip-sample interaction occurs via the presence of liquid films on surfaces, most commonly water. The films are more likely to form under humid conditions. The presence of liquid layers is also dependent on the nature of the surfaces. Some surfaces attract water; these are called hydrophilic surfaces. Some surfaces repel water; these are hydrophobic surfaces. Hydrophilic surfaces have molecules in them that are polar, and they interact with the permanent dipoles of water molecules through the dipole-dipole interactions described in Sect. 6.2.1.

A household example of a moderately hydrophilic surface is aluminum foil. The aluminum oxidizes after it is made. The bonds are polar, with the oxygen atom more negative than the aluminum atom. Non-stick cookware, typically coated with polytetrafluoroethylene (PTFE, Teflon™), is another household example, this time of a hydrophobic surface.

You can directly observe that water is attracted more to aluminum oxide than to the PTFE by placing a small drop of water on each surface. The droplet will be more spread out on the aluminum foil than on the cookware. The droplets assume the shapes that minimize the amount of energy it takes to form their surfaces.

As defined in Sect. 5.3.6, the work of adhesion is the amount of energy required to separate two surfaces from contact to infinity. Each surface can be thought to be responsible for one-half of the work of adhesion. Thus surface energy $\gamma = \varpi/2$. For liquids, the term "surface tension" is preferred, even though its definition is the same as surface energy.

Assuming that your laboratory is sufficiently humid and your tip and sample are sufficiently hydrophilic such that smooth layers of water form on their surfaces, what is responsible for the shape of the curve displayed in Fig. 6.2b?

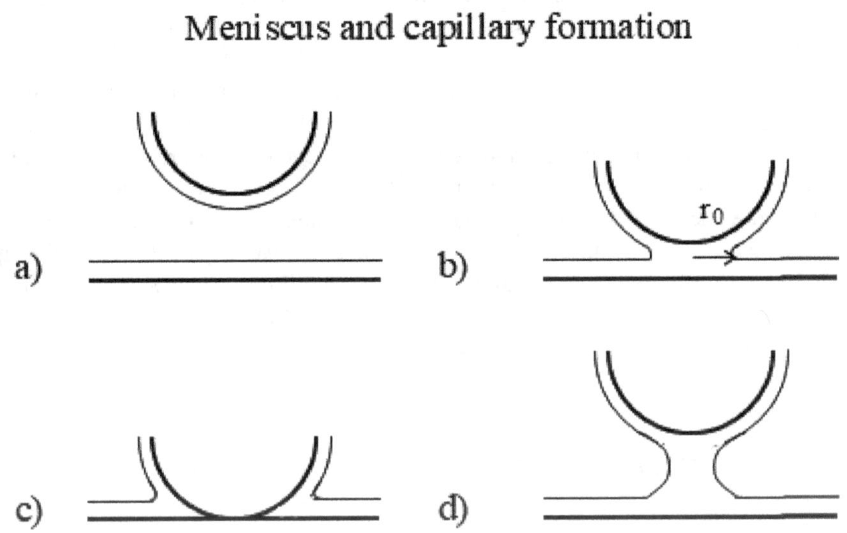

Fig. 6.5 Sketches of the hemispherical tip and flat sample (thick lines) covered with liquid films (thin lines) (a) during the approach of the tip and sample, (b) when the fluid layers make contact and form a meniscus, (c) upon further system compression, and (d) during retraction and the formation of a capillary. In (b), the arrow represents the radius of the initial meniscus, r_0.

6.2.2.1 The hysteretic cycle

Figure 6.5 shows a cycle of tip-sample approach and retraction. In Fig. 6.5a, the tip and sample (thick lines) have not yet made contact. There is an even layer of fluid on each (thin lines). In Fig. 6.5b, the liquid layers have just touched. The amount of surface that has been removed from the liquid layers is approximately πr_0^2, where r_0 is the length of the arrow in Fig. 6.5b. The nonconservative work done on the tip by the liquid during approach is therefore $W_{app} = +2\pi\gamma r_0^2$, and energy goes into accelerating the tip towards the sample and into the potential energy of the bending of the cantilever. Figure 6.5c shows the tip in contact with the sample, which might happen solely because of the minimization of surface area of the fluid, or it might occur with further extension of the scanner. Upon retraction, the fluid forms a capillary, or “liquid bridge", between the tip and sample, as portrayed in Fig. 6.5d. When the capillary breaks, the tip is released, and the remaining potential energy in the cantilever is converted into kinetic energy. The cantilever oscillates until its energy is dissipated into the air; by inference, similar processes also happen within the two fluid layers.

An approximate model for the capillary is to assume that it is cylindrical and maintains its cross-sectional area as it extends. Fluid flows in from the adjacent unperturbed liquid to maintain its area and increase its volume. For a constant radius r_1 and an increase in its length Δl, the work done on the tip due to the increase in surface area during retraction is $W_{ret} = -2\pi\gamma r_1 \Delta l$. In P. 6.6, you will calculate the corresponding force.

Note: the dissipative parts of the force curve are the instabilities; a stable capillary can be extended and contracted reversibly. The total energy lost due to the presence of the liquid films during the approach-retraction cycle is the area of the hysteresis loop of the curve. During the initial instability as the tip accelerates towards the sample, the energy is dissipated into the fluid by the movement of the tip, and if the tip crashes into the sample, it can cause damage as well as heat. When the tip is released from the fluid, the capillary collapses, the liquid sloshes about again, and the cantilever creates turbulence in the air as it oscillates back and forth. An $F(\delta)$ curve will be more representative of the influence of the liquid film than an $F(h)$ curve; the latter includes all dissipation mechanisms.

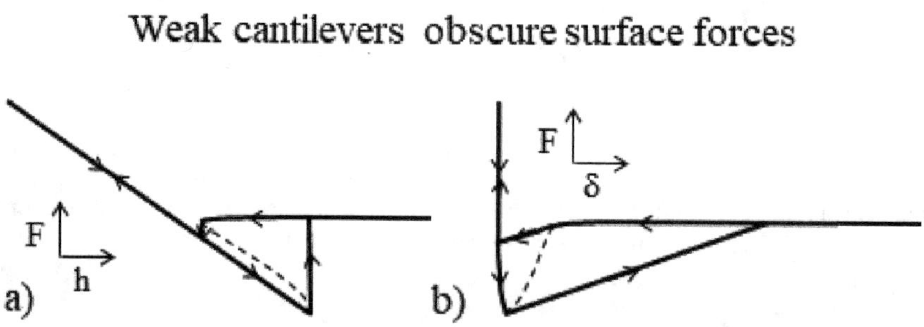

Fig. 6.6 Independent of the nature of the surface forces, the shape of a force curve taken with a weak cantilever will be the same: (a) unprocessed data, $F(h)$, and (b) processed data, $F(\delta)$. (Section 5.2.2 describes how data are processed.) The dashed lines correspond to the unstable equilibrium positions of the tip when the total potential, the sum of the cantilever and van der Waals potentials, is bistable (Sect. 5.1).

> AFM Lesson 26 at youtube.com\AtomicForceMicro
>
> AFM Lesson 26 relates surface forces and spring constants.

6.2.3 Cantilever instabilities and attractive forces

In Sect. 5.1.3, it was noted that weak cantilevers are subject to instabilities. The cantilever can jump over a relevant part of the force curve. Figure 6.6 presents the typical shapes of force curve associated with compliant cantilevers, in both $F(h)$ and $F(\delta)$ formats. The curves have the same shape regardless if the attractive force is van der Waals, meniscus-capillary, or anything else. One simply cannot tell from this shape of curve which attractive forces are operative. All the data are showing you is that your cantilever's spring constant is smaller than the magnitude of the greatest (negative) stiffness of the interaction (p. 98).

From Sect. 5.1.3, when the total potential is bistable, there are two stable positions for the tip, when it is monostable there is just one. The regions of the curve for which the approach and retraction traces overlie each other correspond to the positions for which the total potential is monostable. When the approach and retraction are different, the total potential is bistable. The approach and retraction traces are the stable equilibria. The dashed lines in Fig. 6.6 show the force values corresponding to the unstable equilibria. The word used to describe the force over the bistable region in Fig. 6.6a is "triple-valued."

Now that we have covered the common attractive forces for when the tip and sample are initially not in contact, in the next two sections, our attention turns to when they are in contact.

6.3 Continuum contact mechanics

A simple way to start thinking about when the tip and sample are in contact is to ignore the fact that materials are made of atoms. Rather, for ease of calculation, we consider them to be continuous media. In the next section on molecular dynamics, we will take the opposing view. But the continuum approach leads to tractable equations from which reasonable estimates of the parameters can be made.

6.3.1 Reminder of stress-strain curves

Referring back to Sect. 5.3.1 on stress-strain curves, when a macroscopic sample is stretched, the relationship between stress σ and strain ϵ is simply $\sigma = E\epsilon$ for the linear part of the stress-strain curve, where E is the elastic modulus (the stiffness of a material). It is the elastic modulus that is the usual parameter of interest.

Since stress σ is force F per unit area A, it is straightforward to see that $F = EA\epsilon$. Force calibration for AFMs was presented in Sect. 4.1.2. The challenge now is to understand how to think about the area and strain terms for the volume of the sample that is being compressed underneath a very, very small AFM tip, in contrast to an entire macroscopic sample that is being stretched, as shown in Fig. 5.9.

We can anticipate that the strain will be proportional to the depth of indentation divided by the tip radius or other geometrical factor, and that the exact expression for the area and strain will depend on the tip geometry. We start with the simplest approach, that of Hertzian contact mechanics.

6.3.2 Hertzian mechanics

Hertzian contact mechanics assumes no surface forces, no adhesion, a sphere-flat geometry, that the indentation depth is much less than the tip radius, and that the principal strain is along the surface plane as the tip indents the sample, stretching the surface laterally. Effectively, this means that the indentation depth not only must be much less than the tip radius, but also must be much less than the sample thickness. This set of assumptions is not often fully met in AFM! Nonetheless, the resulting simple equations often provide reasonable estimates for many situations.

The Hertzian expression for force F with indentation modulus K (proportional to elastic modulus E, Eq. 5.15), tip-sample contact radius a (Fig. 5.11), and tip radius R is

$$F = \frac{Ka^3}{R} \approx KR^{1/2}|\delta|^{3/2}, \tag{6.3}$$

where the approximate expression comes from Eq. 5.17, $|\delta| \approx a^2/R$ if $|\delta| \ll R$, with δ the indentation depth. (This book uses a coordinate system of a horizontal x-y plane along the surface of the sample, with the positive z direction normal to the surface, pointing up. As such, for Hertzian mechanics, the depth δ is negative, and hence the absolute value of δ is used.) Force and depth are measured from the AFM's force-curve data, and the tip radius can be calibrated, so it is possible to determine a value for the indentation modulus K.

Comparing the tip-sample interaction spring constant k_i with the cantilever spring constant k_c can help choose the right cantilever for an experiment. Remembering Eqs. 5.7 and 5.8, if $k_i \gg k_c$, the tip does not indent the sample such that information about the sample from the contact region of the force curve cannot be extracted from the data. And if the opposite is true, $k_i \ll k_c$, the cantilever does not bend enough for the force to be measured. To find k_i, one takes the negative derivative of the force with respect to the indentation depth,

$$k_i = -\frac{dF}{d\delta} = \frac{3}{2}K(R|\delta|)^{1/2} = \frac{3}{2}Ka. \tag{6.4}$$

Another useful estimate is for the pressure beneath the tip. Dividing the force by the contact area πa^2,

$$P_H = \frac{Ka^3}{R}\frac{1}{\pi a^2} = \frac{Ka}{\pi R} = \frac{K}{\pi}\left(\frac{|\delta|}{R}\right)^{1/2}, \tag{6.5}$$

which tells us that pressure is greater for stiffer materials at higher indentation depths with smaller tips.

Often, researchers want to experimentally determine the elastic modulus E_s of their samples. Without the Poisson ratio υ_s of the sample, it is not possible to advance beyond the reduced elastic modulus E_r (Eq. 5.14) or the closely related indentation modulus K (Eq. 5.15). (Generally the term $(1-\upsilon_t^2)/E_t$ in both equations is significantly smaller than the corresponding term for the sample and is not a consideration.) As you see from the above equations in this section, it is possible to determine K from within the Hertzian set of assumptions. The rest of the discussion in this chapter focuses on K and allows the reader whether and how to convert from K to E_s.

Finding a good set of experimental parameters can be challenging. A large enough tip radius should be used such that the condition of $|\delta| \ll R$ is met, usually by gluing micron-sized spheres onto the ends of tipless cantilevers. (The unfortunate consequence of large tips is a loss of lateral resolution.) The cantilever spring constant should be chosen to be large enough to indent the sample within the range of the scanner without too much forward sliding on the sample surface (Fig. 5.8a), but small enough to deflect after the tip makes contact with the sample. Pressure should be limited to avoid damaging the tip or sample. Checking your experimental procedure on a known sample with a modulus near that of the sample of interest is highly recommended.

AFM Lesson 27 at `youtube.com\AtomicForceMicro`

AFM Lesson 27 tackles contact mechanics without surface forces.

● In **MacroLab 6**, you determine the reduced modulus of a compliant material. ●

■ The **electronic supplement** to this book contains a spreadsheet that can help you decide what combination of experimental parameters to use to test your material, based on Hertzian mechanics. ■

6.3.3 Other options for contact mechanics

Hertzian contact mechanics provides a good starting point for estimating experimental parameters or to be used for determining modulus. However, frequently the assumption of no surface forces is too restrictive. What other approaches could be useful, without being too complicated?

There are plenty! Each option is based on a certain set of assumptions. Each has advantages and disadvantages. Here, we present three tractable alternatives to Hertzian mechanics: (i) Derjaguin-Muller-Toporov (DMT) mechanics, (ii) Johnson-Kendall-Roberts-Sperling (JKRS) mechanics, and (iii) Burnham-Colton-Pollock (BCP) mechanics. These three models plus the Hertz approach are plotted in Fig. 6.7 as (a) normalized contact radius as a function of normalized depth and (b) normalized force as a function of normalized depth. (The factors by which force and depth are normalized are provided in P. 6.12.) More approaches to contact mechanics can be found in the literature.

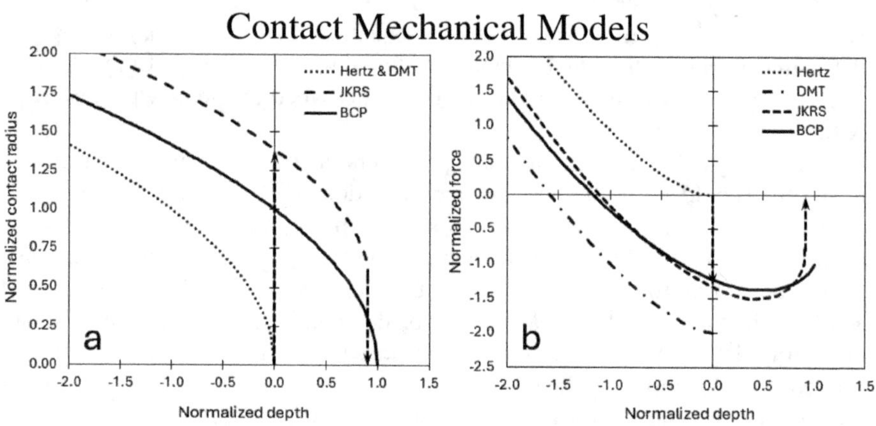

Fig. 6.7 Comparison of the four contact mechanical models described in this book, plotted as (a) normalized contact radius as a function of normalized indentation depth and (b) normalized force as a function of normalized indentation depth. (See P. 6.12 for the normalization factors.) Non-contact forces are omitted for clarity. The Hertz and DMT curves have similar shapes, as do BCP and JKRS. The vertical arrows associated with the JKRS curves indicate how the surfaces will jump into contact upon approach (at zero depth) and experience adhesion hysteresis upon retraction (near +1 depth), whereas the other three models are not inherently hysteretic.

6.3.3.1 DMT mechanics

Derjaguin-Muller-Toporov (DMT) mechanics applies to stiff materials and small tips with non-negligible but low attractive surface forces. Attractive forces are integrated around the periphery of the spherical tip to obtain an attractive term of $-2\pi\varpi R$. The Hertzian contact geometry of $|\delta| \approx a^2/R$ still applies, as the sample is assumed not to bulge out towards the tip during tip-sample approach. As a consequence, the area of contact may be underestimated. Another limitation of these assumptions is that there is a strong discontinuity in slope of the force curve upon contact. The governing equations are

$$F = \frac{Ka^3}{R} - 2\pi\varpi R; \; |\delta| \approx \frac{a^2}{R}. \tag{6.6}$$

In Fig. 6.7, one can observe that the DMT curve has the same shape as the Hertz curve, although the DMT curve is offset negatively because of the attractive forces.

6.3.3.2 JKRS mechanics

Johnson-Kendall-Roberts-Sperling (JKRS) mechanics applies to large tips and compliant materials with strong adhesion within the area that the tip and sample touch. The adhesion is accounted for only within the contact, and it is not operative until tip and sample touch. Hence, JKRS mechanics may underestimate tip-sample attraction upon tip-sample approach.

Upon retraction, the tip pulls on the sample, forming a connective neck before they suddenly lose contact, leading to inherent adhesion hysteresis. The vertical arrows in Fig. 6.7 depict this phenomenon. Contact is made at zero normalized depth, and it is broken near a normalized depth of +1. The equations describing JKRS behavior are

$$F = \frac{Ka^3}{R} - \sqrt{6\pi\varpi K a^3};$$
$$\delta = \frac{2}{3}\sqrt{\frac{6\pi\varpi a}{K}} - \frac{a^2}{R}. \tag{6.7}$$

The contact geometry is not Hertzian; in the δ equation you see that the positive square-root term dominates for small values of the contact radius a, leading to positive values of indentation depth δ: the sample surface is pulled out by the tip. Since the δ equation has terms of both a^2 and $\sqrt{a}$, it requires a fourth-order root when solving the equations.

6.3.3.3 BCP mechanics

Burnham-Colton-Pollock (BCP) mechanics was formulated to address typical AFM experiments where attractive forces are non-negligible (as for DMT and JKRS), yet there are no sharp discontinuities in force or slope of the force curves while making or breaking contact, unlike for DMT and JKRS. Yet like JKRS, the sample surface is noticeably deformable because of attractive forces. Remembering Newton's Third Law, which states that forces come in pairs of equal magnitude but opposite directions, whatever force pulls the tip towards the sample, the sample is pulled towards the tip with the same amount of force. Hence, if the sample has a low modulus or the tip-sample attraction is high (or both!), the sample bulges out. By building a bulge-out distance δ_0 into the equations, the sharp discontinuities of DMT and JKRS mechanics are avoided. BCP's limitation is that at high loads, the force diverges from the Hertzian sphere-flat $|\delta|^{3/2}$ functionality. The BCP equations are

$$F = \frac{Ka^3}{R} - \sqrt{\frac{3\pi\varpi Ka^3}{2}} - \pi\varpi R;$$
$$\delta = \left(\frac{\pi^2\varpi^2 R}{K^2}\right)^{1/3} - \frac{a^2}{R} \equiv \delta_0 - \frac{a^2}{R}. \tag{6.8}$$

In the force equation, the first term is the Hertzian repulsive term common to all the contact mechanics described in this book. The second attractive term accounts for how the surface energy of the tip-sample system varies with depth. The third term represents the attractive forces just before contacting the bulged-out surface. The second term grows in magnitude more quickly than the first term at small values of a, yet at high values of a, the first term dominates. This behavior is a result of the $a^{3/2}$ and a^3 dependencies, respectively. The third term is constant.

In the depth equation, the first term is the bulge-out distance δ_0, and the second term is familiar from the previously described mechanics; for low values of contact radius a, the depth δ is positive. Surprisingly, depending on the ratio of work of adhesion (squared) and tip radius to the indentation modulus (squared), the bulge-out distance can be tens of nanometers. In the case of very compliant materials such as sealants, it can be even greater, high enough such that you can feel the tack with your fingers.

Use the bulge-out distance as a reality check on which contact mechanical model is realistic for your experiment. Whereas others have used an interatomic distance as a cutoff to determine which model might be most appropriate, you can also use the ratio of the maximum indentation depth to the bulge-out distance as a guide. If the ratio is large, it might be reasonable to use the simpler Hertzian mechanics (Sect. 6.3.2).

AFM Lesson 28 at `youtube.com\AtomicForceMicro`

AFM Lesson 28 continues contact mechanics, with surface forces.

★ In **Computer Lab 6**, you observe how sample properties change force curves. ★

■ The **electronic supplement** to this book contains a workbook that can help you decide what combination of experimental parameters to use to test your material. In addition to the previously mentioned spreadsheet based on Hertzian mechanics, there are as well sheets for DMT and BCP mechanics in the same workbook. (JKRS mechanics involves fourth-order roots and is thus omitted for maintaining the simplicity of the workbook.) The BCP sheet calculates the bulge-out distance δ_0 based on your estimates for work of adhesion ϖ, tip radius R, and indentation modulus K, from which you can decide which contact mechanical model might be closest to your experimental situation. ■

6.3.4 Hysteresis everywhere!

Hysteresis in AFM takes many forms. Hysteresis was mentioned in the context of scanner problems in Sect. 4.2.3 and again in Sect. 5.2.3.2. Section 5.1 was devoted to explaining hysteresis from the cantilever through the concepts of potentials, forces, and concavity. And in this chapter's description of JKRS contact mechanics (Sect. 6.3.3.2), sudden adhesion hysteresis—the making and breaking of a connective neck between tip and sample—was described. In this section, another form of adhesion hysteresis via incremental peeling is presented.

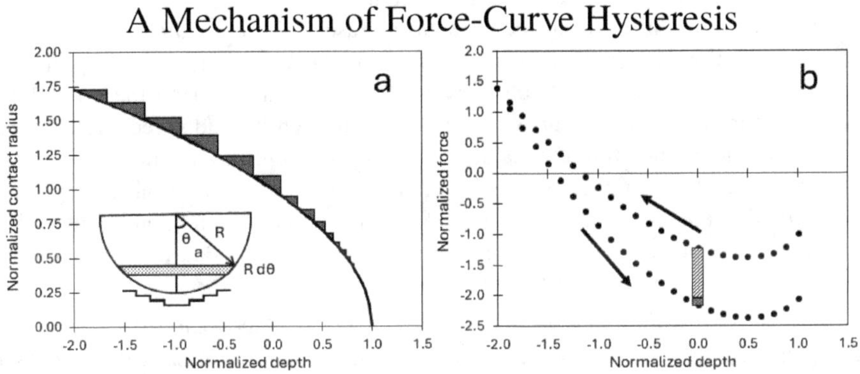

Fig. 6.8 The causes of hysteresis in force curves while in contact are varied. Here one mechanism is illustrated through (a) normalized contact radius and (b) normalized force, both as a function of normalized depth using BCP contact mechanics (Sect. 6.3.3.3). As force is unloaded, the tip and sample separate by means of a sequence of small tugs, represented by the shaded triangles along the curve in (a). Each tug corresponds to a ring of circumference $2\pi a$ and arc length $Rd\theta$ suddenly releasing (inset). (The double staircase sketch immediately below the inset is a reminder that nominally spherical tips are not expected to be perfectly smooth.) The work done in each tug is summed in (b) to account for the unloading (down arrow) curve being below the loading (up arrow) curve. The lightly shaded box at zero depth in (b) represents the sum of work done associated with the shaded triangles in (a) from −2.0 to just below zero depth. The smaller, darkly shaded box just below the lightly shaded box in (b) represents the smaller contribution from the tug at zero depth in (a), where the differential contact area and hence, the work done is much less than at deeper depths.

Instead of adhesion hysteresis occurring abruptly at contact, as for JKRS mechanics, we can think about it happening step-by-step during the unloading process—a peeling of the two surfaces from each other. Figure 6.8 illustrates the process, where normalized contact radius and normalized force for BCP mechanics are plotted as a function of normalized separation or indentation depth in (a) & (b), respectively. The inset of (a) represents a nominally spherical tip that is unlikely to be perfectly smooth (double staircase below the tip). A circumference of $2\pi a$ and height $Rd\theta$ is shaded; its differential area is $2\pi aRd\theta$. The tip and sample lose contact gradually, each increment corresponding to loss of contact at a succession of ever smaller circumferences, and each requiring an additional increment of force over a vertical displacement in order to separate.

On the curve of the contact radius in (a), the horizontal parts of the shaded triangles (or steps, if you prefer) of decreasing area represent the contact radius initially remaining the same as for a small change in depth as force is unloaded, then suddenly releasing down to the nominal contact area (vertical sides of triangles). While the triangular area in (a) corresponds to a distance squared (radius times depth), the work done by each successive tug of force can be calculated based on the separation force per unit area, the differential area of the circumference, the change of depth during the retraction of the scanner, and their relative orientations (P. 6.16). This separation force per unit area is in addition to van der Waals or other long-range forces; it is a consequence of contact and can be thought of as the force required to break chemical bonds that formed during contact.

Work is done during each incremental release while unloading the force. As a result, the unloading part of a force curve will be below the loading part of the curve, as seen in Fig. 6.8b. Each increment of movement during the unloading is subject to an additional amount of work done. The lightly shaded rectangle between the loading and unloading curves at zero depth in (b) represents the work done up until zero depth; it is not to scale with the triangles in (a). The small darkly shaded box directly beneath it represents the work done at zero depth. Consistent with the changing area of the shaded triangles in (a), the darkly shaded box in (b) is small compared to the sum of the amount of work done up to that point.

Two other possible causes of hysteresis in the contact region of a force curve are plasticity and viscoelasticity of the sample; the tip is assumed to withstand the pressures during loading and behave elastically. Plasticity refers to permanent damage. If plasticity is present, a subsequent force curve on the same position at the same load is less likely to show as much hysteresis, and an image of the area after indenting will show where a hole has been created in the sample.

If instead the hysteresis is caused by viscoelasticity—the time dependent response of the sample—the hysteresis will be a function of the loading rate. The faster the force curve is acquired, the lower the hysteresis. An image of the area after indenting will look similar to an initial image taken before you started force-curve acquisition, if the time scale of the recovery allows the sample to return to its original state between force-curve and image acquisition.

If you believe the hysteresis is dominated by plasticity or viscoelasticity, then the determination of indentation modulus should be done from the unloading section of the force curve, where the inelastic effects have taken place during loading. If on the other hand you believe that the hysteresis is dominated by adhesive peeling, then fit your contact mechanical model to the loading part of the curve. Then adhesive peeling—not included in any contact mechanical model that I am aware of—does not falsely contribute to the modulus value. At the time of this writing in 2024, commercially available fitting programs do not conduct this analysis for you.

Hysteresis is the formal word for a forward and a reverse process not being the same, be it extension and retraction of a scanner, cantilever instabilities occurring at different z-positions of the scanner, formation and breaking of a connective neck, gradual formation and peeling as the tip-sample interface changes area, or the plasticity or viscoelasticity of the sample. There are many sources of hysteresis in AFM!

6.3.5 What about different tip geometries?

So far in this text we have assumed that the end of the tip at the end of the cantilever is spherical. Let us now consider alternatives. In this section, we shall assume low surface forces, that is, Hertzian contact mechanics.

The functional dependence of force on depth depends on how quickly the perimeter of the contacting area increases with depth. We can think of the perimeter as determining an effective spring constant. Multiplying the spring constant times the depth yields the magnitude of the force.

We consider three cases, as portrayed in Fig. 6.9: (a) A flat-ended punch does not change its contact area with depth. The force is linear with depth and dependent on the indentation modulus K and contact radius a, much like how a spring or cantilever force is linearly dependent on its spring constant k_c. (b) For a spherical tip, the perimeter $2\pi a \approx 2\pi(R|\delta|)^{1/2}$ increases as the square root of depth, giving a force that depends on the 3/2 power of depth. (c) A cone's perimeter increases linearly with depth, such that the force goes as the square of the depth. The exact equations are shown below. The new variable is the half-angle θ of the cone.

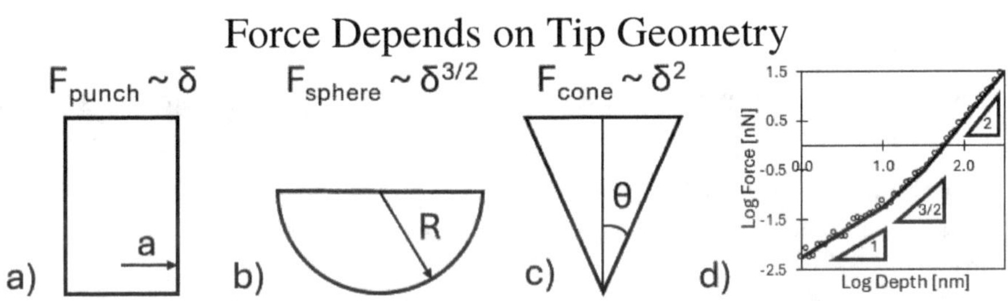

Fig. 6.9 The geometry of the tip affects the depth dependence of the force: (a) the force for a flat-ended punch of radius a is linearly dependent on the indentation depth; (b) the force for a spherical tip is dependent on the 3/2 power of the depth; and (c) the force for a conical tip depends on the square of the depth. (d) A log-log plot can reveal the effective shape of the tip via the slope.

$$F_{punch} = \frac{3Ka}{2}|\delta|,$$

$$F_{sphere} = KR^{1/2}|\delta|^{3/2}, \tag{6.9}$$

$$F_{cone} = \frac{3K\tan\theta}{2\pi}\delta^2.$$

It can be difficult to establish the exact shape of the tip. Here is where log-log plots can be useful. Taking the logarithm to the base 10 of the equations above results in

$$\log(F_{punch}) = \log\left(\frac{3Ka}{2}\right) + \log(|\delta|),$$

$$\log(F_{sphere}) = \log(KR^{1/2}) + \frac{3}{2}\log(|\delta|), \tag{6.10}$$

$$\log(F_{cone}) = \log\left(\frac{3K\tan\theta}{2\pi}\right) + 2\log(|\delta|).$$

If plotted as Log Force as a function of Log Depth, the data will have a slope that reflects the tip geometry: 1 for a punch, 3/2 for a sphere, and 2 for a cone. The y-intercept is the first term, from which K can be determined, as long as a, R, or θ are known, respective to the governing equation. It sometimes happens that all three slopes are seen, as represented in Fig. 6.9d. In this case, there are three equations from which three unknowns may be determined.

The reader should be aware that logarithms may not be taken of negative numbers, such that depths must always be positive. The reader should also be attentive to the units. For example, if depth is given in millimeters and force in milliNewtons, the calculated indentation modulus will be in kiloPascals. Other combinations can be practiced in P. 6.17 & 6.18.

6.3.6 Recommendations for force-curve acquisition and analysis

How should you go about preparing for, acquiring, and analyzing force curves? Here are my recommendations, along with where in this textbook more information is provided:

- If you are primarily interested in the force of adhesion between your tip and sample, your challenges are fewer than if you are interested in the modulus or the work done during force-curve acquisition. However, do be mindful of how the chemistry and geometry of the tip (and possibly the sample) can change because of contact. *Sect. 6.2.*

- If you are interested in more than the force of adhesion, ensure that you have a stiff enough cantilever so as to avoid cantilever instabilities. Remember that instabilities can be due to not only the strength of the tip-sample attractive forces, but also the size and shape of the tip and the sample topography. *Sects. 5.1, 5.3.4 & 6.2.1.3.*

- Based on the anticipated attractive forces and modulus, decide which contact-mechanical model best fits your situation. Choose experimental parameters to indent your sample without excessive strain and to minimize horizontal sliding, yet maintain detectable flexion of the cantilever. *Sect. 6.3, Appendix B.4.*

- Calibrate the spring constant and deflection sensitivity of your cantilever. Confirm that your scanner is calibrated. Calibrate your tip shape and size. Confirm that your force curves follow the depth dependence of your assumed tip geometry. *Sects. 4.1.2, 4.1.2.2, 4.1.3, & 6.3.5.*

• If your force curves exhibit hysteresis in the contact or noncontact regions, consider whether the contact hysteresis might originate from plasticity, viscoelasticity, or adhesive peeling. For the noncontact regions, consider the possibility of fluid films. For both contact and noncontact, consider possible changes in tip chemistry or geometry and changes in sample topography. *Sects. 5.3.4, 6.2.2, 6.3.4, & 6.3.5.*

• If you are interested in surface forces—attraction and adhesion—use contact mechanics to help you decide where contact begins and ends. Distinguish between attraction (at initial tip-sample contact) and adhesion (as they separate) in your analysis. Attraction contributes to adhesion, yet adhesion might also include an abrupt JKRS-like release or an adhesive peeling effect due to tip-sample contact. *Sects. 6.2 & 6.3.*

• If you are interested in modulus, decide if any hysteresis in your force curves is primarily due to plasticity, viscoelasticity, or adhesive peeling. If plasticity or viscoelasticity is dominant, the contact mechanical model should be fit to the unloading section of the curve. If adhesive peeling is dominant, the model should be fit to the loading section of the curve. *Sect. 6.3.4.*

In **Instrument Lab 6,** you demonstrate your competence with AFM.

Researchers for whom these continuum contact mechanical models are useful might also want to learn about molecular dynamics, an alternative way to study nanoscale contacts using a computational approach. An introduction to molecular dynamics is supplied in the next section.

6.4 Molecular dynamics

Continuum contact mechanical theories (Sect. 6.3) do not consider the atomic nature of matter. Molecular dynamics does. Tips and samples are composed of atoms in computer simulations. By approaching the study of contacts in this manner, different insights have been obtained, e.g., the transfer of material from the sample to the tip.

The equations for continuum contact mechanics are based on a set of assumptions. So are molecular dynamics simulations. In this section, we first examine the assumption that all atoms are particles, then discuss how atoms move within a potential created by their interaction with nearby atoms, followed by an overview of how a molecular dynamics simulation is done, and finally we give a comparison between the continuum and molecular dynamics approaches.

6.4.1 Do objects act as particles or waves?

You might have heard of wave-particle duality, the quantum-mechanical phenomenon that entities such as electrons and photons behave either as waves or particles, but not both at the same time. Whether the entity behaves as one or the other depends on the circumstances.

For example, electrons are often introduced to students as point particles. When we think of current flowing in a wire, we think of the number of particle-like electrons passing by per second. In contrast, scanning tunneling microscopy (STM) is entirely dependent on the wavelike nature of electrons, in that the STM tip detects the electrons that leak out of the surface (Sect. 3.2.2).

Another example is that of light. When we think of light hitting a solar panel, we think of a particle—a photon—bringing energy to be absorbed by the panel and converted into electricity. But when we observe the colors in a soap bubble, we consider light as a wave; the colors are a result of the constructive and destructive interference between the light passing through the bubble and the light reflected within the bubble.

Interestingly, while electrons and photons exhibit this "on-off" switching between wavelike and particle-like behavior, the likelihood of an object to exhibit wave-particle duality to do so is *not* a switch. Rather, it is a gradual transition dependent on an object's mass and velocity. Its velocity is in turn often dependent on its temperature. To evaluate an object's tendency to behave in this way, the *de Broglie wavelength* λ is used:

$$\lambda = \frac{h}{m\vartheta}, \tag{6.11}$$

where $h = 6.626 \times 10^{-34}$ J·s, m is mass, and ϑ is the magnitude of its velocity. Low-mass objects moving slowly have long enough de Broglie wavelengths to affect neighboring objects. For electrons and atoms, velocity is usually a function of temperature according to

$$\frac{3}{2}k_B\Upsilon = \frac{1}{2}m\langle\vartheta^2\rangle, \tag{6.12}$$

with Υ temperature in Kelvin. The equation represents how thermal energy (the left-hand side) can manifest as kinetic energy (the right-hand side). The thermal energy is evenly divided among the three components of velocity, ϑ_x, ϑ_y, and ϑ_z; this thought is known as the *equipartition theorem*. Referring back to Chap. 4, we used the equipartition theorem with respect to a cantilever's one-dimensional potential energy, Eq. 4.7, for the calibration of its spring constant (Sect. 4.1.2.1).

If we solve Eq. 6.12 for temperature and substitute it into Eq. 6.11 to calculate the de Broglie wavelength, we find that at room temperature the wavelengths of electrons, hydrogen atoms, and helium atoms are larger than their diameters. In contrast, the de Broglie wavelengths of all other atoms and molecules are smaller than their diameters. Although an arbitrary boundary, many scientists are comfortable with thinking of the less massive particles as waves and the more massive ones as particles. Nonetheless, molecular dynamics calculations treat hydrogen and helium as particles at room temperature. Let us now consider how the particles move due to thermal energy.

AFM Lesson 29 at youtube.com\AtomicForceMicro

AFM Lesson 29 introduces wave-particle duality.

6.4.2 How do trapped atoms move?

Every atom has thermal energy and moves as a consequence. If an atom is trapped in a potential created by its interaction with other atoms, the atom vibrates within the potential. We consider again the Lennard-Jones potential (U_{LJ}, Eq. 6.1) as a function of interatomic distance r in order to develop an understanding of thermal vibration;

$$U_{LJ} = U_0 \left[\left(\frac{r_0}{r}\right)^{12} - 2\left(\frac{r_0}{r}\right)^{6} \right]. \tag{6.13}$$

In P. 6.23, you are asked to derive the force, spring constant, equilibrium position, and the potential at equilibrium for U_{LJ}. You shall show that the equilibrium position is r_0, and that $-U_0$ is the potential at equilibrium.

Turning now to Fig. 6.10, a typical Lennard-Jones potential U_{LJ} is plotted in electron volts for potential and nanometers for interatomic distance r. The horizontal line labeled E_{tot} represents an atom's total energy, the sum of its kinetic and potential energies; $E_{tot} = K_E + U_{LJ}$. The energy is exchanged between kinetic and potential energies, with its sum always equal to E_{tot}. For this situation, the total energy is its thermal energy. The atom oscillates back and forth in the potential. It turns around at r_1 and r_2, where its kinetic energy has dropped to zero, down from its maximum kinetic energy at the minimum of the potential curve.

The amount that an atom oscillates depends not only on its total (thermal) energy, but also on the depth and shape of the potential that it is in. For a one-dimensional Lennard-Jones potential, the thermal energy can be all kinetic or all potential according to

$$\frac{1}{2}k_B\Upsilon = \frac{1}{2}m\langle\vartheta^2\rangle = \frac{1}{2}k_{LJ}\langle\Delta r^2\rangle, \tag{6.14}$$

where Δr is the root-mean-square vibrational amplitude. In P. 6.23b, you are asked to find k_{LJ} from the Lennard-Jones potential. Using your answer along with the above equation, one finds

$$\Delta r = \sqrt{\frac{k_B\Upsilon r_0^2}{72U_0}}. \tag{6.15}$$

Thus the vibrational amplitude is affected by Υ, r_0, and U_0.

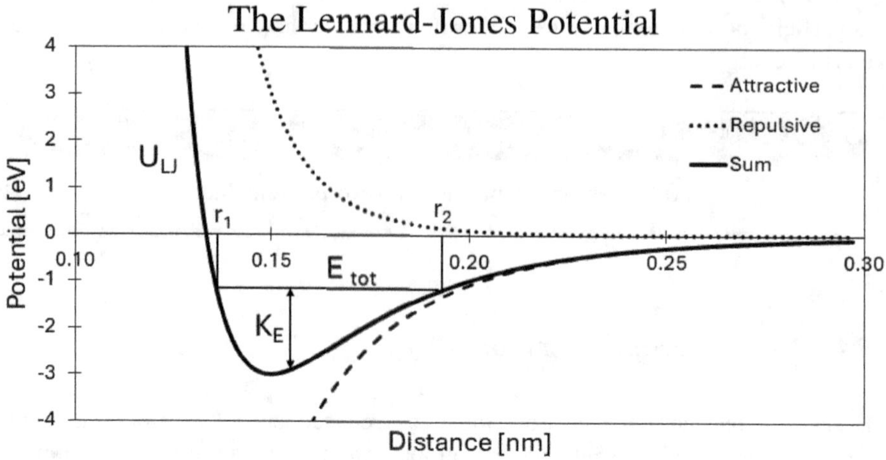

Fig. 6.10 A typical Lennard-Jones potential U_{LJ} in electron volts as a function of interatomic distance r in nanometers, formed by summing attractive and repulsive contributions proportional to $-1/r^6$ and $+1/r^{12}$, respectively (Eq. 6.13). The kinetic energy of an atom trapped in the potential is K_E, and its total energy (horizontal line) is $E_{tot} = K_E + U_{LJ}$. The turning points r_1 and r_2 correspond to where $K_E = 0$.

The Lennard-Jones potential was one of the first potentials to be applied in molecular dynamics simulations. However, it is a simplification, in that the potential is radially symmetric. Radially symmetric potentials only fully apply to noble gas solids. In contrast, more sophisticated and accurate modern potentials reflect the way in which atoms bind to each other along specific directions. We covered the simple Lennard-Jones potential here to present the concept of thermal motion, which is evident in all molecular dynamics simulations. We proceed to examine how molecular dynamics works.

★ In the first part of **Computer Lab 7**, you investigate thermal motion. ★

6.4.3 How does molecular dynamics work?

In a molecular dynamics computer simulation, "atoms" are arranged as they would be in the materials of interest. A representation is sketched in Fig. 6.11, in which the spherical atoms are in the shapes of a tip and a sample. Some atoms are "fixed", others "free". The fixed atoms are allowed to vibrate thermally, but they are not allowed to move from their original positions relative to neighboring atoms. The free atoms not only vibrate thermally, they are allowed to move relative to their neighbors. The layers of fixed atoms are far from the tip-sample interface; the free ones are near the interface. The temperature and relative position of the tip and sample are controlled through the fixed atoms. By regulating the thermal vibrational amplitude of the fixed atoms, the free atoms can be cooled or heated. By moving the set of fixed atoms of the tip or sample, the relative positions of the tip and sample can be changed.

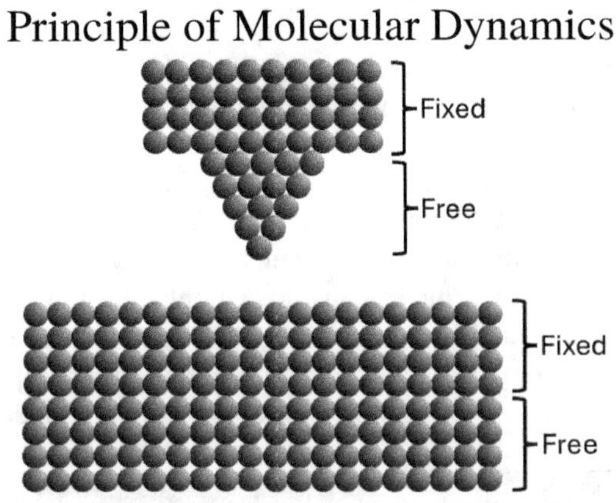

Fig. 6.11 Representation of how a molecular dynamics simulation for AFM is set up, with rows of atoms (spheres) in the shape of a tip (above) and sample (below). The atoms far from the central area between tip and sample are fixed in place, although they are allowed to vibrate with thermal energy. The atoms in the central area are not only allowed to vibrate due to temperature, but also leave original positions if the forces acting upon them are great enough. The fixed atoms can be regulated for their temperature and moved during the simulation, thereby causing changes in temperature and position of the free atoms.

A molecular dynamics simulation accounts for all of the potentials acting between all of the atoms. The negative derivative of potential is force. All the forces acting on an atom are vectorially summed, and the acceleration is found from Newton's Second Law, $\Sigma \mathbf{F} = m\mathbf{a}$. From acceleration, the velocity and position of the atom is found for a small time step, on the order of femtoseconds. Each velocity and position of every atom is found for a given time step, then the entire process is repeated. Any change in the positions of the fixed atoms is programmed to occur after the system equilibrates following the previous position change of the fixed atoms. The behavior of the free atoms is then observed. They sometimes move from their original positions, occasionally even ending up on the tip if they started within the sample, or from the tip to the sample.

The number of calculations for a full molecular dynamics simulation might reach billions, depending on the number of atoms and the overall simulation time. As a consequence, some molecular dynamics simulations might run on even a powerful computer for days, and even then might only simulate a microsecond of real time. A consequence of the short time of molecular dynamics simulations is that longer lasting phenomena in materials, such as how mechanical or optical excitations turn into heat through a process called phonon decay, is not always captured in the simulations.

If short computational times are the practical route for learning about a system, two approaches are often taken to limit computational times.The first approach is that the number of atoms is usually not much more than the number that can be expected to reveal the phenomenon of interest. The clever approach to making the sample seem bigger is to use periodic boundary conditions: whatever atomic movements there might be on the right edge of the simulated tip or sample influence their respective left edges. Similarly, the front and back edges influence each other in the same way. The hard work of such a simulation lies in constructing potentials that accurately reflect the materials involved, such that the researcher can be confident that the simulation mirrors reality as well as possible.

The second approach is that the electrons of the "atom," instead of being in the orbitals that represent the chemical bonds, are represented by a sea of uniform negative charge. This simplification simulates non-magnetic metals reasonably well, in that some electrons of a metal can move freely throughout it, yet no van der Waals forces are included. The size of the "atom" in a movie made from a molecular dynamics simulation remains that of an atom with its full complement of electrons. To simulate magnetism, electron spin would have to be included in the calculation.

AFM Lesson 30 at `youtube.com\AtomicForceMicro`

AFM Lesson 30 examines molecular dynamics.

★ In **Computer Lab 7**, you also play with a molecular dynamics simulation. ★

6.4.4 Continuum mechanics versus molecular dynamics

Now that the basic concepts and some simpler equations of both continuum contact mechanics and molecular dynamics simulations have been presented, it is time to compare them.

Continuum contact mechanics assumes that matter is uniform: the complexity of tracking individual atoms or electrons is put aside. Instead, the contact mechanical theories discussed in Sect. 6.3 are sets of tractable equations. The equations can be used by non-specialists to provide a guide for preparing for an experiment to help choose experimental parameters or to calculate the sample's modulus. However, plasticity of the tip or sample or material transfer between tip and sample are not captured in continuum mechanics. You can think of continuum mechanics as a "top-down" approach.

In comparison, molecular dynamics simulation is a "bottom-up" approach, in that individual atoms are tracked. There can be millions of equations; this field is dominated by specialists who devote their careers to improving the simulations of how materials behave in nanoscale contacts. The results have included plasticity, material transfer from sample to tip or *vice versa*, or even the formation of atomic wires as a tip is retracted from a soft, adhesive metal. While the movies made from molecular dynamics simulations are both entertaining and insightful, the researcher should bear in mind that many simulations ignore the behavior of electrons, meaning that no magnetic or van der Waals forces are accounted for in the results.

Congratulations! You made it to the end of the material for the AFM courses taught at Worcester Polytechnic Institute and ETH Zurich...yet there is still one more chapter.

Chapter 7 is more mathematically advanced, intended to help you understand the behavior of oscillating cantilevers, such as in non-contact or intermittent-contact modes and dynamic force curves. If you have had a course in differential equations and are familiar with complex numbers, you should be able to follow the development. The chapter ends with an explanation of why there are sometimes abrupt changes in the amplitude and phase of the cantilever, which can complicate interpretation of your data.

6.5 Learning objectives, further reading, and problems

! Learning objectives for Chapter 6

6.1 Distinguish between attraction and adhesion.
6.2 Describe the origin of the van der Waals potential.
6.3 Understand how the van der Waals potential takes different functional forms for different geometries.
6.4 Interpret the features of a force curve due to meniscus and capillary formation.
6.5 Sketch how the van der Waals and meniscus-capillary forces manifest themselves in force curves taken with weak and stiff cantilevers.
6.6 Identify the principal features of a force vs depth curve.
6.7 Be able to manipulate and use the Hertz equations.
6.8 Give reasons why there are other contact mechanical models.
6.9 Distinguish between plasticity and adhesion hysteresis.
6.10 Discuss the advantages and disadvantages of the continuum versus molecular dynamics approaches.

Further Reading

✓**Book.** Johnson, K. L. (1987). Contact mechanics. Cambridge university press.

✓**History.** Burnham, N. A., & Colton, R. J. (1989). Measuring the nanomechanical properties and surface forces of materials using an atomic force microscope. Journal of Vacuum Science & Technology A: Vacuum, Surfaces, and Films, 7(4), 2906–2913.

✓**Book.** Israelachvili, J. N. (2011). Intermolecular and surface forces. Academic press.

✓**Tutorial.** Stan, G., & King, S. W. (2020). Atomic force microscopy for nanoscale mechanical property characterization. Journal of Vacuum Science & Technology B, 38(6).

? Problems for Chapter 6

6.1 (a) Sketch Fig.6.3b and label the four force vectors with the charges that create each of them. (b) Also sketch the corresponding four forces acting on the induced dipole at **r** and label them. (c) Indicate the directions of the net forces acting on the original and induced dipoles.

6.2 (a) Make explicit the steps between the interatomic potential $U_{a-a} = -B/r^6$ and the atom-sample potential $U_{a-s} = -\pi \varrho_s B/r^3$, where ϱ_s is the number density of dipoles in the sample. (b) Integrate again to demonstrate that you can find the potential per unit area for a small flat of area A, $U_{f-s}/A = -H/12\pi\delta^2$. (c) Starting from U_{a-s}, write out the steps to U_{t-s}.

6.3 Derive the van der Waals force between a large hemispherical tip a short distance away from a large smooth flat sample, starting from the tip-sample potential, $U_{t-s} = -HR/6\delta$.

6.4 Find $H_{\omega=0}$ and $H_{\omega>0}$ (Eqs. 6.2) in meV for a tip-sample combination of silicon nitride and fused quartz, (a) with air as the intervening medium and (b) with acetal ($C_6H_{14}O_2$) as the medium. The numerical values of dielectric constant and refractive indices are $\varepsilon_{Si_3N_4} = 7.5$, $n_{Si_3N_4} = 2.05$; $\varepsilon_{air} \simeq 1.0\, n_{air} \simeq 1.00$; $\varepsilon_{quartz} = 3.8$, $n_{quartz} = 1.448$; $\varepsilon_{acetal} = 3.8$, $n_{acetal} = 1.38$. Assume an excitation frequency of $\omega_e = 2.0 \times 10^{16}$ rad/s and a temperature of 300 K. (c) Describe in words the relative strengths of the permanent and fluctuating dipoles in the Hamaker constant for the two situations.

6.5 In this problem, you will explore how H can be positive (attractive force), negative (repulsive force), or zero, based on Eqs. 6.2. (a) The smallest possible values of the dielectric constant and index of refraction are for vacuum, $\varepsilon_{vac} = 1$, $n_{vac} = 1$. For air, the values are very close to one, $\varepsilon_{air} \simeq 1.0$ and $n_{air} \simeq 1.0$. For any two bodies in air or vacuum, is H positive or negative? Why? (b) For any medium, if the two bodies are of the same substance ($\varepsilon_t = \varepsilon_s$, $n_t = n_s$), is H greater or less than zero? Explain. (c) If the two bodies and the intervening medium are all different substances, under what circumstances is H negative? (d) What can one do to make van der Waals forces become zero?

6.6 (a) Calculate the force during retraction by a cylindrical capillary on the tip, $W_{ret} = -2\pi\gamma r_1 \Delta l$, which assumes a constant cross-sectional area. (b) For the parameters $\gamma_{H_2O} = 72$ mJ/m^2, $r_1 = 10$ nm, what is the numerical value of the force in nN? (Better fits to experimental data can be achieved by assuming constant volume of the capillary and by closer attention to how surface tension is affected by vapor pressure. However, the mathematics become more complicated.)

6.7 Show that Eqs. 6.3–6.5 are dimensionally correct.

6.8 For an indentation modulus K of 10 kPa, a tip radius R of 10 μm, and an indentation depth $|\delta|$ of 1 μm, calculate the (a) force F and (b) interaction spring constant k_i. If the cantilever spring constant k_c is 0.2 N/m, what are (c) the deflection d of the cantilever and (d) the z-motion of the scanner when tip and sample are in contact?

6.9 For an indentation modulus K of 1 GPa, a tip radius R of 1 μm, and an indentation depth $|\delta|$ of 10 nm, calculate the (a) force F and (b) interaction spring constant k_i. If the cantilever spring constant k_c is 200 N/m, what are (c) the deflection d of the cantilever and (d) the z-motion of the scanner when tip and sample are in contact?

6.10 (a) From Eqs. 6.3–6.5, find dF/da, $dF/d|\delta|$, dk_i/da, $dk_i/d|\delta|$, dP_H/da, $dP_H/d|\delta|$. (b) Which ones are constants? (c) Which ones go to infinity at small values of a or $|\delta|$?

6.11 Show that Eqs. 6.6–6.8 are dimensionally correct.

6.12 In Fig. 6.7, four contact mechanical models are displayed as normalized contact radius and normalized force as a function of normalized indentation depth. Using the normalization relationships $\bar{F} = F/(\pi\varpi R)$, $\bar{A} = a/(\pi\varpi R^2/K)^{1/3}$, $\bar{\Delta} = \delta/(\pi^2\varpi^2 R/K^2)^{1/3}$, find $\bar{F}$ and $\bar{\Delta}$ in normalized form for (a) Hertz, (b) DMT, (c) BCP, and (d) JKRS.

6.13 Use the BCP equations to find (a) the pressure $P = F/\pi a^2$, where a is the contact radius of the area between the tip and sample, (b) the rate at which the force F changes with a, dF/da, and (c) compare how the pressure and force rate behave if a is very small.

6.14 Use the BCP equations to derive (a) the rate at which the force changes with contact radius, dF/da, (b) the rate at which the indentation depth changes with contact radius, $d|\delta|/da$, and (c) the rate at which force changes with indentation depth $dF/d|\delta|$ in terms of the variables K, a, R, ϖ, for the case when tip and sample are in contact. (d) How is $dF/d|\delta|$ related to the interaction stiffness k_i? (e) If firm contact between the tip and sample corresponds to negative values of $|\delta|$ by virtue of the choice of coordinate system, is k_i less than or greater than zero?

6.15 In Fig. 6.8, the concept of work done by adhesive peeling is presented. Is the work done positive or negative? Why? *Hint:* Look at the geometry of the inset of Fig. 6.8a. Which way does the force $\mathbf{f}$ of the sample on the tip point relative to the displacement $\Delta\delta$ during unloading?

6.16 In Fig. 6.8a, shaded triangles are shown, representing how the contact radius can stay the same over a small increment of depth. The triangular areas vary according to the amount of work done at each release of differential contact area around the circumference. (a) What is the differential area, dA? (b) What is the differential work done, dW, for a differential area? Express your answer in terms of the separation force per unit area $\mathbf{f}_A$, the displacement $\Delta\delta$, R, θ, and trigonometric functions. (c) Explain in words why the areas of the triangles change with depth and at what angle θ would you expect the largest area. (d) Integrate your expression for dW from θ_0 to 0 for the total work done upon tip-sample separation.

6.17 You want to make a logarithm (to the base 10) plot of your force-curve data. You know that you have to keep a consistent set of units in order to find the indentation modulus K of your sample. (a) If you use nanometers for the absolute value of the indentation depth and tip radius, and nanoNewtons for the force, what units result for the indentation modulus? (b) The straight-line portion of the log-log plot with a slope of 3/2 intersects the y-axis at a value of +1.0. If the radius of the tip is 100 nm, what is the indentation modulus K? (c) You notice that the initial portion of the curve has a slope of 1, intersecting the y-axis at +1.2. What is the contact radius a of the tip's flat-ended sphere?

6.18 You want to make a logarithm (to the base 10) plot of your force-curve data. You know that you have to keep a consistent set of units in order to find the indentation modulus K of your sample. (a) If you use micrometers for the absolute value of the indentation depth and microNewtons for the force, what units result for the indentation modulus for a nominally conical tip? (b) You are curious to see a force curve that reflects the conical shape of your tip, despite your hesitations about too much strain for the Hertz equations to apply. The straight-line portion of the log-log plot with a slope of 2.0 intersects the y-axis at a value of -2.0. If your tip has a half-angle θ of 25°, what is the indentation modulus K? (c) The initial part of the curve has a slope of 3/2, and it intersects the y-axis at -1.8. Assuming that your value of K is valid, what is the tip radius?

6.19 An $R = 100$ nm tip is indented such that the tip-sample contact radius $a = 5.0$ nm. Its cross-sectional contact area is πa^2. (a) In contrast, what is the surface area in contact with the sample? (*Hint:* Start with the differential area $dA = 2\pi aRd\theta$, make a substitution, and integrate from θ_0 to 0.) (b) If an atom takes up $(0.25\ \text{nm})^2$ on the surface, how many pairs of atoms are on either side of the interface between tip and sample? (c) If the hysteresis between the loading and unloading curves within the contact portion of a force curve equals 1.0 fJ, what is the energy in eV required to break the bond(s) between an interfacial pair of atoms?

6.20 (a) What is your de Broglie wavelength λ when you are at rest? (b) What is it when you are moving at 5.0 km/hr, a typical walking speed? (c) Is it possible to detect your walking-speed λ with today's technology?

6.21 Calculate the de Broglie wavelength λ in nanometers of (a) an electron, (b) a hydrogen atom, and (c) a helium atom, all at 77 Kelvin. (d) Look up their diameters. What are they? (e) By our definition of particles and waves, which of (a)–(c) would be considered particles and which waves?

6.22 Calculate the de Broglie wavelength λ in picometers of (a) a proton at 77 Kelvin, (b) a carbon atom at 77 Kelvin, (c) a carbon atom at 300 Kelvin, and (d) a silicon atom at 300 Kelvin. (e) Look up the diameters of a proton, carbon atom, and silicon atom. What are they? (f) By our definition of particles and waves, which of (a)–(d) would be considered particles and which waves?

6.23 (a) Derive the Lennard-Jones force, $F_{LJ}(r)$, and (b) spring constant, $k_{LJ}(r)$, from the Lennard-Jones potential, Eq. 6.1. (c) Find the equilibrium position, r_{eq}. (d) What are $U_{LJ}(r_{eq})$ and $k_{LJ}(r_{eq})$? (e) Sketch where r_{eq} and $-U_{LJ}(r_{eq})$ appear on Fig. 6.10.

6.24 Figure 6.10 was made with values $U_0 = 3.0$ eV and $r_0 = 0.15$ nm. What is the mean-square vibrational amplitude, Δr, in picometers at (a) 300 Kelvin? (b) 77 Kelvin? *Hint:* Use your answer to P. 6.23b.

6.25 A helium atom is trapped in the Lennard-Jones potential of Fig. 6.10, which has values of $U_0 = 3.0$ eV and $r_0 = 0.15$ nm. The temperature is 293 Kelvin. (a) What is the atom's root-mean-square velocity, $\sqrt{\langle \vartheta^2 \rangle}$? (b) If the helium atom is replaced by an argon atom, what is the new root-mean-square velocity? (c) If the temperature is reduced to 4.0 Kelvin, what is the new value of $\sqrt{\langle \vartheta^2 \rangle}$ for the argon atom?

6.26 A good research AFM might with a weak cantilever installed might have a force resolution of 10 pN. (a) If you were able to apply 10 pN of force to just one oxygen molecule, O_2, what would be the resulting acceleration? (b) If the molecule was intially at rest, for how long should the force be applied to reach a speed of 200 m/s (a typical vibrational speed at room temperature)? Answer in picoseconds.

6.27 Look online for a publication that uses force-curve analysis or molecular dynamics to help address the research question. Send the link to your instructor. Be prepared to discuss the paper in class with a five-minute summary.

6.28 Pre-Lab Question. You are planning an experiment to determine the indentation modulus K of a material of your choosing or that your instructor has assigned to you. You want to select the appropriate cantilever spring constant, tip radius, and maximum indentation depth to do so. Use the workbook in the electronic supplement of this book to determine which model (Hertz, DMT, BCP) and which set of experimental parameters might be best for your research. Be prepared to defend your intended plan in five minutes to your class.

6.29 Lab question. If a sample were given to you for which you knew that its two components (e.g., a matrix with inclusions) were very different in properties but you didn't know which was which, what AFM techniques could you use to decide between them?

6.30 *Challenge Problem.* Here you will derive the inverse-sixth power dependence of the Lennard-Jones potential on distance. (a) What is the electric field vector $\mathbf{E}_{1on2} = \sum_i \frac{1}{4\pi\varepsilon_0}\frac{q_i}{r_i^2}\hat{r}_i$ at the polarizable atom positioned at $\mathbf{r}$ in Fig. 6.3a? (b) The electric field $\mathbf{E}_{1on2}$ induces a dipole q_2a_2 at $\mathbf{r}$ in Fig. 6.3b. The strength of the induced dipole depends on the atom's polarizability χ and the magnitude of $\mathbf{E}_{1on2}$. Thus, $q_2a_2 = \chi E_{1on2}$. Solve for q_2. (c) By vectorially summing the forces, find the net force acting upon the original dipole from the induced dipole, $\mathbf{F}_{2on1}$. Assume $a_1 = a_2$. This answer depends on both q_1 and q_2 (d) Use the answer to (b) to express $\mathbf{F}_{2on1}$ in terms of q_1 only. (e) In the limit of $r \gg a$, use the binomial theorem, $(1+\xi)^n \approx 1 + n\xi$, where $\xi \ll 1$, to simplify your expression for $\mathbf{E}$. (f) By Newton's Third Law, what is the net force acting on the induced dipole, $\mathbf{F}_{1on2}$? (g) Find the negative integral of $\mathbf{F}_{2on1}$, that is, the U_6 potential.

6.31 *Challenge Problem.* It is instructive to solve for significant points on the normalized force vs. depth plot of Fig. 6.7. Starting with your answers to P. 6.12, make a table where the columns are: Model, $\bar{\Delta}(\bar{F}=0)$, $\bar{F}(\bar{\Delta}=0)$, $(\bar{\Delta}_{min}, \bar{F}_{min})$, $(\bar{\Delta}_{MB}, \bar{F}_{MB})$, and the rows are the DMT, BCP, and JKRS values for the column headings. The "min" subscript refers to the minimum value of force, and the "MB" subscript refers to where contact is made and broken.

Chapter 7
Oscillator models for AFM

Abstract In the realm of physics and engineering, oscillators play a pivotal role, particularly in the context of Atomic Force Microscopy (AFM). The oscillation of the cantilever is a key aspect, with the simplest case being the free-hanging cantilever subject only to gravity—a scenario modeled by the simple harmonic oscillator. In noncontact mode, small forces modify this model, leading to the perturbed harmonic oscillator. Both models exhibit linear behavior, where the cantilever's response is directly proportional to the excitation. However, these linear models fall short in describing all AFM modes, notably in intermittent-contact mode, where tip-sample interactions become highly nonlinear. Traditional linear relationships break down, requiring researchers to turn to numerical simulations for accurate representation. This chapter introduces a more intuitive approach to understanding the complexities of intermittent contact. The motivation behind this more mathematical chapter is twofold: to enhance your comprehension of dynamic modes (noncontact and intermittent contact) and to provide practical insights for applications such as calibrating phase images, determining cantilever quality factors, and optimizing intermittent-contact mode imaging. While we will delve into some mathematical intricacies, higher modes of cantilever oscillation will not be covered, which are more complicated still.

7.1 Linear oscillators

7.1.1 The model for the simple harmonic oscillator

Figure 7.1 shows the model for the simple harmonic oscillator (SHO). The z-axis points up, away from the sample surface. The cantilever holder is at position $h(t)$; the reflective side of the cantilever above the tip is at $c(t)$. The cantilever is represented by a spring of stiffness k_c, a dashpot of damping coefficient b_c, and a point mass m. One can apply a load (force) $F(t)$ directly to the tip (e.g. magnetically, if the cantilever is coated with a magnetic material), or jiggle the cantilever holder $h(t)$. In earlier chapters, c, h, and F changed slowly enough such that at any instant one could think of them as static. This condition is called quasistatic. Now, in this chapter, we must think of our variables as functions of time, i.e. dynamic.

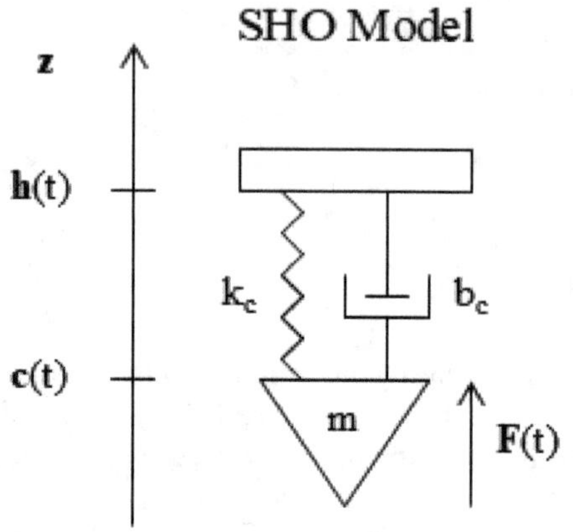

Fig. 7.1 Model used to understand the behavior of a simple harmonic oscillator. A spring of stiffness k_c, a dashpot with damping coefficient b_c, and a point mass m comprise the cantilever, which is subject to applied load $F(t)$ or oscillation of the holder of its base $h(t)$. The back of the cantilever's tip is at $c(t)$.

The spring constant k_c has been discussed and used in Sects. 4.1.2 and 5.2. A dashpot is a device like the shock absorber in your car that absorbs energy. The bigger the value of b_c, the more it absorbs energy. The damping of a cantilever in air or water is predominantly due to the turbulence it creates as it moves through the medium, whereas in vacuum, the significantly smaller damping is due to more complicated mechanisms within the material. The model assumes that all of the mass of the cantilever is concentrated at its tip. This proves to be an adequate assumption for the situations that we will consider here, but is not valid at frequencies much higher than that at which the cantilever would naturally vibrate, that is, its frequency of resonance.

Newton's Second Law states that the sum of the forces acting on a body equals its mass times its resulting acceleration. The body of interest is the mass of the tip. The forces acting upon it are the force due to the spring (proportional to the difference in position of its two ends), the force due to the dashpot (proportional to the difference in speed between its two ends), and the applied load. In Fig. 7.1, the equilibrium position of the cantilever is at $h(t)$, where the spring and dashpot are neither compressed nor extended. The position of the tip, as drawn, reflects extension. All the forces are acting so as to move the mass up. Newton's Second Law is

$$m\ddot{c} = b_c(\dot{h} - \dot{c}) + k_c(h - c) + F. \tag{7.1}$$

The dots on top of the variable mean time derivatives. One dot is the first derivative with respect to time (speed), two dots the second derivative with respect to time (acceleration). For convenience, we drop the explicit vector and time dependencies, e.g. $c(t)$ is written as just plain c. The static solution, that is, when nothing is moving ($\ddot{c} = \dot{c} = \dot{h} = 0$), should be familiar to you; $F = -k_c(h - c)$. Rearranging Eq. 7.1,

$$m\ddot{c} + b_c\dot{c} + k_c c = b_c\dot{h} + k_c h + F. \tag{7.2}$$

This formulation of Newton's Second Law is called the equation of motion for the cantilever. It determines the behavior of the tip of the cantilever. Solving it for c as a function of all the different parameters, our objective, can be done efficiently with complex numbers.

7.1.2 Solving for the tip's motion

This section makes use of complex numbers in order to find the solutions to the equation of motion for the cantilever. The results should be useful to you, even if you are unfamiliar with the mathematics.

7.1.2.1 Complex numbers

Imagine a vector with its tail at the origin of a two-dimensional coordinate system, with its head rotating in a counter-clockwise direction. The arbitrary function $\mathbf{f}$ corresponding to this vector has the form $\mathbf{f} = r\,(\exp i\omega t)\ \hat{r}$, where r is the magnitude (length) of the vector, ω its radial frequency, and t time. The unit vector $\hat{r}$, which indicates the direction of the vector, is related to the cartesian-coordinate unit vectors $\hat{\mathbf{x}}$ and $\hat{\mathbf{y}}$ by $\hat{r} = \cos\omega t\ \hat{\mathbf{x}} + \sin\omega t\ \hat{\mathbf{y}}$. Since $\exp i\omega t = \cos\omega t + i\sin\omega t$,

$$\mathbf{f} = r(\cos\omega t + i\sin\omega t)\,\hat{r}. \tag{7.3}$$

A complex function is formed from real and imaginary parts, in other words,

$$\mathbf{f} = Re(\mathbf{f})\,\hat{\mathbf{x}} + iIm(\mathbf{f})\,\hat{\mathbf{y}}. \tag{7.4}$$

Comparing Eqs. 7.3 and 7.4, the real part is represented by $Re(\mathbf{f}) = r\cos\omega t = x$ and the imaginary part by $Im(\mathbf{f}) = r\sin\omega t = y$.

Figure 7.2 summarizes the essentials of complex numbers. The x-axis is the real axis, the y-axis the imaginary axis. The vector of length r at angle ωt from the x-axis rotates counter-clockwise. The vector can be broken into its x- and y-components (real and imaginary components, respectively) by trignometry. Similarly, familiar relations can be used to find the magnitude r and the angle ωt.

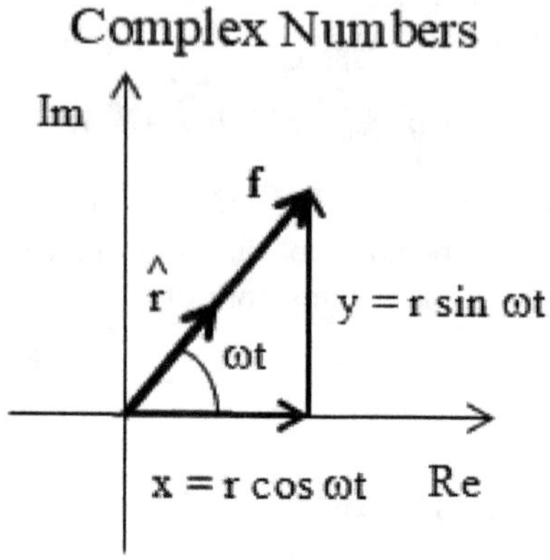

Fig. 7.2 An arbitrary vector **f** of magnitude r in the complex plane, where the x-axis is the real axis and the y-axis the imaginary axis. The head of the vector rotates in a counter-clockwise direction at radial frequency ω. Its tail remains at the origin. The unit vector $\hat{r}$ gives the direction in which **f** points at any instant in time. Trigonometry is used to find its real and imaginary components.

$$\begin{aligned} r &= \sqrt{x^2 + y^2} \\ &= \sqrt{[Re(\mathbf{f})]^2 + [Im(\mathbf{f})]^2}, \\ \tan \omega t &= \frac{y}{x} \\ &= \frac{Im(\mathbf{f})}{Re(\mathbf{f})}. \end{aligned} \tag{7.5}$$

Thus armed, we return to the problem of finding the position of the cantilever as a function of time. The cantilever holder position $h(t)$ and the applied load $F(t)$ are complex functions of the form $r\ (\exp i\omega t)\ \hat{r}$. The cantilever also oscillates as $\exp i\omega t$, but we allow it to be phase-shifted, that is, its response can be advanced or delayed in time with respect to the excitation, h or F. Mathematically we then write the function as $c = c_1 \exp i\,(\omega t + \theta)\ \hat{r}$, where θ is the phase angle. The subscript 1 is a reminder that we are now solving for an oscillatory amplitude. Technically, in all of the formulae that have been written thus far, we should have been using the subscript 0 to represent the static positions—it was dropped for simplicity. Finding the position of the cantilever as a function of time means solving for both the amplitude c_1 and the phase θ. One of our assumptions—a valid one—is that the cantilever moves at the same frequency as the excitation, i.e. is harmonic with whatever drives it.

The first steps in solving for the amplitude and phase are to substitute $c_1 \exp i(\omega t + \theta)$, $h_1 \exp i\omega t$, and $F_1 \exp i\omega t$ into Eq. 7.2, taking the time derivatives, finding $c_1 (\exp i\theta)\ \hat{r}$, and multiplying numerator and denominator by the complex conjugate of the denominator, keeping the denominator real. (If $\mathbf{f} = Re(\mathbf{f})\,\hat{\mathbf{x}} + iIm(\mathbf{f})\,\hat{\mathbf{y}}$, then the complex conjugate of $\mathbf{f}$ is $\mathbf{f}^* = Re(\mathbf{f})\,\hat{\mathbf{x}} - iIm(\mathbf{f})\,\hat{\mathbf{y}}$.) This process results in

$$\begin{aligned} c_1 (\exp i\theta)\ \hat{r} &= \left(\frac{(k_c - \omega^2 m)\,\hat{\mathbf{x}} - i\,(\omega b_c)\,\hat{\mathbf{y}}}{(k_c - m\omega^2)^2 + (\omega b_c)^2} \right) F_1 \\ &+ \left(\frac{\left[k_c\,(k_c - \omega^2 m) + (\omega b_c)^2\right]\hat{\mathbf{x}} - i\,(\omega^3 m b_c)\,\hat{\mathbf{y}}}{(k_c - m\omega^2)^2 + (\omega b_c)^2} \right) h_1. \\ &\equiv J_F F_1 + J_h k_c h_1 \end{aligned} \tag{7.6}$$

Don't let this mess scare you. The left-hand side contains our sought-for variables, c_1 and θ. The right-hand side has two terms, one a function of F_1 and the other a function of h_1. One or the other of these is zero at any given time for most experiments. The denominators on the right-hand side are identical, and the numerators are broken into their real ($\hat{\mathbf{x}}$) and imaginary ($\hat{\mathbf{y}}$) parts. J_F and J_h are defined as complex compliances; compliance has the units of meters per Newton, the inverse of spring constant.

When the base of the cantilever does not move, yet a load is applied directly to the free end of the cantilever, typically by means of a magnetic field (for magnetic cantilevers), then this choice of operation is called force control. Its opposite, displacement control, describes the situation when no load is applied to the cantilever's free end and the base of the cantilever is moved. You will see that the cantilever behaves differently depending upon how it is controlled and depending on the frequency at which it is excited.

7.1.2.2 Force-control solutions

To solve for the cantilever's amplitude c_1 and phase θ under force control, the real and imaginary parts of Eq. 7.6 are substituted into Eqs. 7.5 for the case of $h_1 = 0$. The result is

$$\begin{aligned} c_1 &= \frac{F_1}{\sqrt{(k_c - \omega^2 m)^2 + (\omega b_c)^2}}, \\ \tan\theta &= \frac{-\omega b_c}{(k_c - \omega^2 m)}. \end{aligned} \tag{7.7}$$

These are the parameters that we use in $c_1 \exp i(\omega t + \theta)\ \hat{r}$ to describe the motion of the tip end of the cantilever. On our complex coordinate system, the solution looks as in Fig. 7.3. The excitation is F_1, with the real part of $J_F F_1$ in Eq. 7.6 in phase with it. The imaginary part of $J_F F_1$ lags by 90°. The real and imaginary parts are the components of $c(t)$ in Eq. 7.6, with magnitude c_1, lagging F_1 by the angle θ.

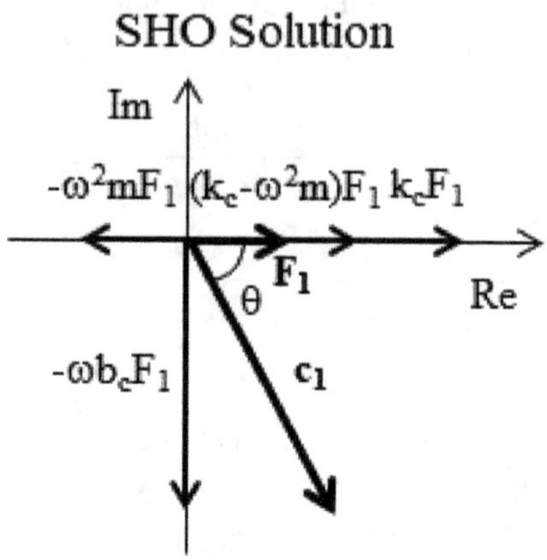

Fig. 7.3 The solution to the simple harmonic oscillator (SHO) equation of motion as represented by vectors in a complex coordinate system. The excitation $\mathbf{F}_1$ and response $\mathbf{c}_1$ are shown as the thickest arrows, the contributions to the response as medium arrows, and the axes as narrow arrows. The response $\mathbf{c}_1$ is the hypoteneuse of the triangle formed by the sum of the real contributions, proportional to $(k_c - \omega^2 m)\, F_1$, and the imaginary contribution, proportional to $-\omega b_c F_1$.

The sketch in Fig. 7.3 is like a snapshot taken in time. Do not forget that the time-dependent form of $c(t)$ is $c_1 \exp i\,(\omega t + \theta)\,\hat{r}$ and that the driving force F_1 is also oscillatory. Imagine that all of the vectors rotate in a counter-clockwise fashion, with all of their tails at the origin, at a rate of ω radians per second. The amplitudes c_1 and F_1 and their phase difference θ are maintained as long as all of the parameters remain constant.

The form of Eqs. 7.7 is good for some circumstances and awkward for others. An alternative presentation comes from normalizing them. This means that not only do the right- and left-hand sides become dimensionless (no units), but they depend on dimensionless variables. To carry this process out, we define two useful quantities, $\omega_k^2 \equiv k_c/m$ and $Q \equiv \sqrt{k_c m}/b_c = k_c/\omega_k b_c$ (to be discussed shortly) and rewrite the unnormalized force-control equations above as

$$\frac{k_c c_1}{F_1} = \frac{1}{\sqrt{\left[1 - \left(\frac{\omega}{\omega_k}\right)^2\right]^2 + \left(\frac{\omega}{\omega_k Q}\right)^2}},$$

$$\tan\theta = \frac{-\omega}{\omega_k Q\left[1 - \left(\frac{\omega}{\omega_k}\right)^2\right]}. \tag{7.8}$$

7.1.2.3 Displacement-control solutions

Cantilever motion is more commonly instigated by an oscillating cantilever holder than by a time-dependent applied load. To solve for the cantilever's amplitude c_1 and phase θ under displacement control, the real and imaginary parts of Eq. 7.6 are substituted into Eqs. 7.5 for the case of $F_1 = 0$. We now examine the consequences of holding F_1 equal to zero while h_1 is nonzero. Equations 7.6 and 7.5 result in

$$c_1 = h_1 \sqrt{\frac{\left[k_c(k_c - m\omega^2) + (\omega b_c)^2\right]^2 + \left[m\omega^3 b_c\right]^2}{\left[\left(k_c - \omega^2 m\right)^2 + (\omega b_c)^2\right]^2}},$$

$$\tan\theta = \frac{-\omega^3 m b_c}{k_c\left(k_c - \omega^2 m\right) + (\omega b_c)^2}, \tag{7.9}$$

for the unnormalized solution, and for the normalized solution

$$\frac{c_1}{h_1} = \sqrt{\frac{\left[1 - \left(\frac{\omega}{\omega_k}\right)^2 + \left(\frac{\omega}{\omega_k Q}\right)^2\right]^2 + \left[\frac{\omega^3}{\omega_k^3 Q}\right]^2}{\left[\left[1 - \left(\frac{\omega}{\omega_k}\right)^2\right]^2 + \left(\frac{\omega}{\omega_k Q}\right)^2\right]^2}},$$

$$\tan\theta = \frac{-\omega^3}{\omega_k^3 Q\left[1 - \left(\frac{\omega}{\omega_k}\right)^2 + \left(\frac{\omega}{\omega_k Q}\right)^2\right]}. \tag{7.10}$$

7.1.3 Making sense out of the equations

7.1.3.1 SHO amplitude and phase

Let us now try to make sense out of Eqs. 7.7 and 7.8, the force-control solutions for the motion of the cantilever's tip. Later, in Sect. 7.1.3.3, we shall compare the force and displacement control equations.

For force control, Fig. 7.4a shows the amplitude c_1 as a function of radial frequency ω (Eqs. 7.7), both plotted on logarithmic (base 10) scales. Equations 7.8 appear in Figure 7.4b, similarly with logarithmic scales. Logarithmic scales are used when the numbers vary over a large range because taking logarithms makes small values out of large ones, and thus a wider range fits onto a small scale. The right axes correspond to phase angle, which is a linear scale, since the range extends only from 0° to −180°. The negative phase angle stems from the minus sign in the $\tan\theta$ expressions above. It tells us that the response lags the excitation.

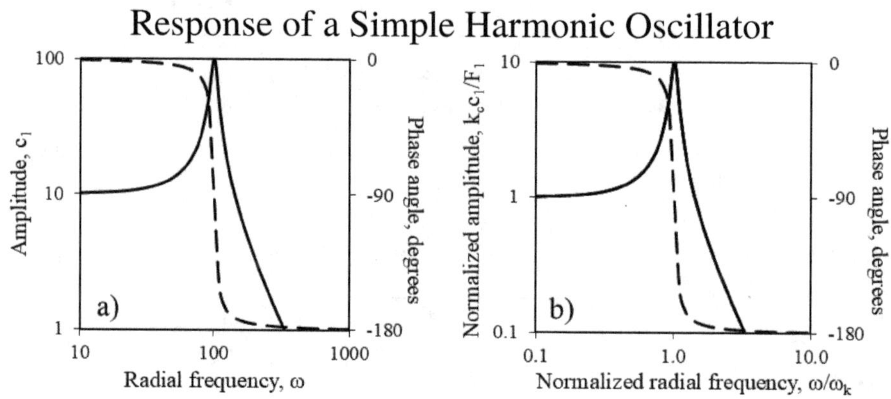

Fig. 7.4 The unnormalized and normalized amplitude (solid lines, left-hand scales) and phase angle (dashed lines, right-hand scales) of a simple harmonic oscillator with $Q = 10$. In (a), amplitude is presented as a function of radial frequency, whereas in (b), normalized amplitude is plotted as a function of normalized radial frequency. Normalization does not affect the shape of the curves, only the axes' values. The phase angle does not change because the expression for $\tan\theta$ is already dimensionless.

First examining the behavior of c_1 (Eqs. 7.7) as a function of frequency, in the example of Fig. 7.4(a) we see that the low-frequency value of c_1 is 10. At higher frequencies c_1 passes through a maximum and then falls to small values. The maximum in amplitude occurs near (but not at) ω_k, the frequency at which the kinetic energy in the cantilever is a maximum. This is called the natural or kinetic resonance frequency, and it equals $2\pi\nu_k$. Parameters that affect the magnitude of c_1 at low frequencies are the amplitude of the applied load F_1 and the spring constant k_c. Near resonance, the load F_1, the kinetic resonance ω_k, and the damping coefficient b_c determine the peak height. The phase equals zero at low frequencies and falls to −90° at kinetic resonance, then continues to −180° at high frequencies.

In P. 7.5 you shall take the low- and high-frequency limits and the at-resonance values of Eqs. 7.7 and 7.8. Substitute $\omega = 0$ into the equations for the low-frequency limit, $\omega = \infty$ for the high-frequency limit, and $\omega = \omega_k = \sqrt{k_c/m}$ for the resonance case. By playing with a spring, you can check to see if you believe that the equations are a good reflection of reality. Look for both amplitude and phase behavior as a function of frequency.

Figure 7.4b displays Eqs. 7.8. This time the normalized amplitude $k_c c_1/F_1$ and the phase angle θ are plotted as a function of normalized radial frequency ω/ω_k. Note that the shapes of the curves have not changed, but the axes have. Now kinetic resonance occurs at $\omega/\omega_k = 1$, the low-frequency amplitude = 1, the height of kinetic resonance is Q (per P. 7.5), and as before, the amplitude in the high-frequency limit falls to zero. Normalization can help reveal behavior; all curves based on Eqs. 7.8 with the same Q have the same shape.

7.1.3.2 Determining quality factor and phase angle

What is Q? It is called the quality factor, an important characteristic of an oscillator, and is equal to $\sqrt{k_c m}/b_c$ or, equivalently, $k_c/\omega_k b_c$. In Fig. 7.4b, the maximum height of the peak is marginally greater than Q. High stiffness, low kinetic resonance frequency, or low damping all mean high Q and therefore a taller resonance peak. The quality factor also affects the shape of the peak. Low-Q peaks are broader than high-Q ones. Additionally, the quality factor determines the oscillation decay time. If an oscillator is drawn away from its equilibrium position and simply released, it can oscillate numerous times before coming to rest. Its amplitude decays exponentially in time, following the mathematical relation $\exp(-o\pi/Q)$, where o is the number of oscillations. The factor Q/π acts as a time constant. If o equals Q/π, the amplitude has decayed to $1/e$ (37%) of its original value. For $o = Q$, the amplitude has decayed to 4.3% of its starting magnitude. Thus one can say that Q is approximately the number of oscillations necessary for the oscillator to stabilize after a change in its excitation.

The quality factor must be determined for some cantilever stiffness calibration methods. If one were to know k_c, ω_k, and b_c from independent methods, one could calculate Q. Or, if one could obtain the ratio of the amplitudes at low frequency and at kinetic resonance, Q could be found. In practice, however, b_c is not measured directly, and the low-frequency end of the plot is often obscured by noise. Here is a practical way to find Q. The frequency of maximum amplitude ω_a, the amplitude resonance, is divided by the frequency difference between the points on the curve where the amplitude has dropped by a factor of $1/\sqrt{2}$, $\omega_+ - \omega_-$. Hence,

$$Q \approx \frac{\omega_a}{\omega_+ - \omega_-}. \tag{7.11}$$

Now we explain why this relation works. You can use calculus to differentiate the amplitude formula of Eqs. 7.8 with respect to the quantity ω/ω_k. Set the result equal to zero and solve for ω/ω_k. You will find that the maximum amplitude occurs at a frequency of $\omega_a = \omega_k\sqrt{1 - 1/2Q^2}$. Substituting this frequency back into the amplitude equation yields the maximum amplitude of $Q/\sqrt{1 - 1/4Q^2}$. (If Q is large, $\omega_a \approx \omega_k$ and the maximum amplitude is very close to Q.) Coming down in amplitude to a value of $Q/\sqrt{2\left(1 - 1/4Q^2\right)}$, you can solve for the corresponding frequencies by noting that the square of the inverse of the amplitude equation is quadratic in $(\omega/\omega_k)^2$ and applying the quadratic formula to solve for $(\omega/\omega_k)^2$ and thus ω/ω_k. There are two roots, one corresponding to the positive sign and one corresponding to the negative sign. They are $\omega_+ \approx \omega_a(1 + 1/2Q)$ and $\omega_- \approx \omega_a(1 - 1/2Q)$. The difference between ω_+ and ω_- equals ω_a/Q.

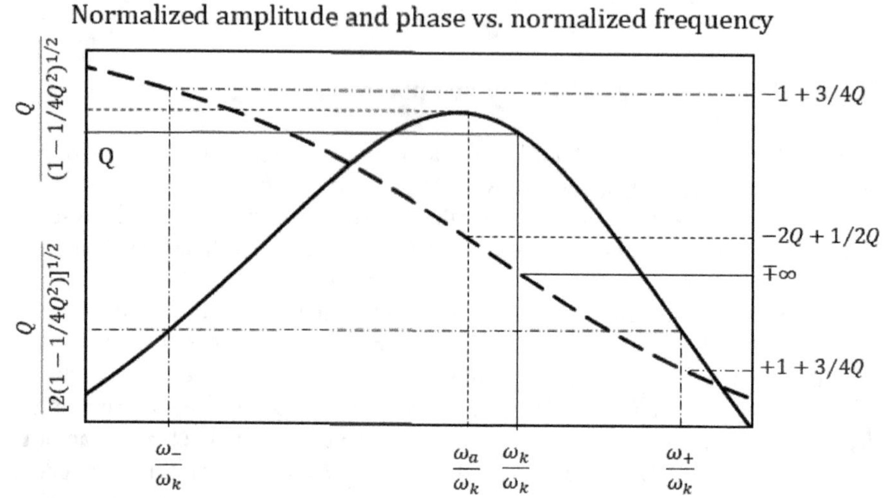

Fig. 7.5 Normalized amplitude (solid line) and tangent of the phase angle (dashed line) as a function of normalized radial frequency for a force-controlled simple harmonic oscillator with $Q = 2$. For frequencies of particular interest, the normalized amplitude and tangent of the phase angle are indicated on the graphs in terms of Q on the left- and right-hand scales, respectively. The thin solid lines connect the values associated with the normalized kinetic resonance ω_k/ω_k. The thin dashed lines indicate the values for normalized amplitude resonance ω_a/ω_k, and the thin dot-dash lines for ω_-/ω_k and ω_+/ω_k.

Figure 7.5 displays these points on a spectrum of normalized amplitude (left-hand scale) as a function of normalized radial frequency for an oscillator with $Q = 2$. The peak is rather broad and slightly asymmetric in comparison to the peaks in Fig. 7.4, where the quality factor was ten. As Q increases, the asymmetry decreases. The peak also looks different because a linear scale was chosen due to the limited ranges of interest of the amplitude and frequency. The phase response is also shown, where the $\tan\theta$ values for ω_-/ω_k, ω_a/ω_k, ω_k/ω_k, and ω_+/ω_k are given on the right-hand scale, although beware that terms smaller than $1/2Q$ have been dropped. You can verify that these values are correct by substituting the expressions for the frequencies into the formula for phase angle in Eqs. 7.8. Knowing the correct phase angles corresponding to distinct frequencies gives you a means of verifying if your microscope is generating proper phase readings. Often the accuracy of phase angle in AFMs is poor.

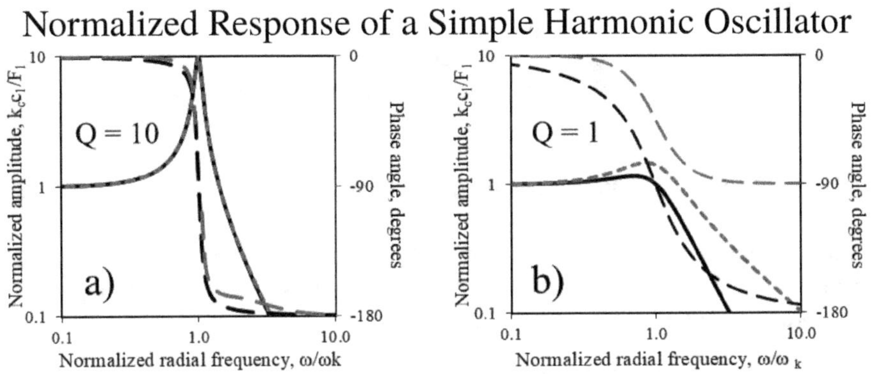

Fig. 7.6 Normalized amplitude (solid and short-dashed lines, left-hand scale) and phase angle (long dashed lines, right-hand scale) as a function of normalized radial frequency for (a) $Q = 10$ and (b) $Q = 1$. The force-control solutions are in black and the displacement-control solutions are in gray. The lower the quality factor, the more distinct the force- and displacement-control responses become, especially above resonance.

7.1.3.3 Comparing force and displacement control

Displacement control's expressions, Eqs. 7.9 and 7.10, are more complicated than the force-control expressions, Eqs. 7.7 and 7.8. One notices the same denominators in the amplitude expressions and somewhat different terms in the tangent denominator and numerators. To what extent do they reveal the same behavior?

If, for example, $Q \gg \omega/\omega_k$ in the normalized equations, the displacement-control ($F_1 = 0$) equations evolve to being close to the force-control ($h_1 = 0$) ones. A spectral comparison is that of Fig. 7.6, where normalized amplitude and phase angle for both displacement and force control are plotted as a function of normalized frequency for $Q = 10$ in (a) and $Q = 1$ in (b). Notice that the shapes of the peaks are dependent on Q. Only the phase angles at frequencies above resonance deviate slightly from each other for $Q = 10$. For $Q = 1$, force and displacement control are significantly different, especially at frequencies above resonance.

The lesson of Fig. 7.6 is that the simpler force-control equations are reasonable approximations to the displacement control case if the quality factor is high. In general, quality factors in air are thirty or higher (in the thousands for vacuum) and below ten when the cantilever is immersed in fluids. In a dense medium, then, the displacement-control equations must be used if the experiment is displacement-controlled.

7.1.4 The electrical equivalent of a cantilever

Physicists take great satisfaction from seeing multiple ways in which the same mathematical formalism can be applied. The equations governing the behavior of a mechanical force-controlled simple harmonic oscillator as presented in Sect. 7.1.3

apply to electrical circuits with an inductor, resistor, and capacitor in series. For those of you who are comfortable with complex numbers, differential equations, and ac circuits, this section is meant as supplemental material.

7.1.4.1 Mechanical and electrical equivalence

Assume that an ac voltage source $V(t)$ drives an inductor L, a resistor R, and a capacitor C in series. The equation of motion for this circuit is fully equivalent to our force-controlled cantilever,

$$L\ddot{q} + R\dot{q} + \frac{q}{C} = V(t). \tag{7.12}$$

Using the same approach as in Sect. 7.1.2 for finding the solution,

$$q_1 \left(\exp i\theta\right) \hat{r} = \left(\frac{\left(\frac{1}{C} - \omega^2 L\right) \hat{\mathbf{x}} - i\left(\omega R\right) \hat{\mathbf{y}}}{\left(\frac{1}{C} - \omega^2 L\right)^2 + (\omega R)^2} \right) V_1 \equiv \frac{Y}{\omega} V_1, \tag{7.13}$$

where Y is the admittance of the circuit. As a reminder,

$$c_1 \left(\exp i\theta\right) \hat{r} = \left(\frac{\left(k_c - \omega^2 m\right) \hat{\mathbf{x}} - i\left(\omega b_c\right) \hat{\mathbf{y}}}{\left(k_c - \omega^2 m\right)^2 + (\omega b_c)^2} \right) F_1 \equiv J_F F_1, \tag{7.14}$$

so, by comparing Eq. 7.14 with Eq. 7.13 one can establish the following analogs between the electrical and mechanical variables:

electrical	q_1	V_1	$1/C$	L	R	Y/ω
mechanical	c_1	F_1	k_c	m	b_c	J_F

You should now be able to find the expression for the amplitude and phase of the charqe by two separate routes — firstly, by using the real and imaginary parts of Eq. 7.13 and the definitions of amplitude and phase, Eqs. 7.5; and secondly, by substituting the electrical equivalent of the mechanical variables into Eqs. 7.7.

A word of caution is that this simple analogy between a series LRC circuit and a cantilever cannot be pushed too far. The analogs between variables listed above do not apply to a displacement-controlled cantilever, and one must turn to a more complicated electrical configuration.

7.1.4.2 Dissipated power

Nonetheless, it is interesting to calculate the power dissipated by the series *LRC* circuit, and, by implication, the cantilever. Dissipated power is $(ReI)(ReV)$ (or force and speed); current is the derivative with respect to time of charge, $I(t) = i\omega q(t)$. Since current and voltage vary with time, the time average of power over one cycle of oscillation is our goal, denoted by brackets, as in $\langle P \rangle$. Taking the time average means integrating in time over the interval of one period $(2\pi/\omega)$, then dividing by the length of that interval (the period, again). Mathematically, the time average looks like

$$\langle P \rangle = \frac{\omega}{2\pi} \int_0^{\frac{2\pi}{\omega}} (ReI)(ReV)dt \tag{7.15}$$

$$= \frac{\omega}{2\pi} \int_0^{\frac{2\pi}{\omega}} Re\left[i\omega q_1 \exp i(\omega t + \theta)\right] Re\left[V_1 \exp i(\omega t)\right] dt \tag{7.16}$$

$$= \frac{\omega}{2\pi} \int_0^{\frac{2\pi}{\omega}} \left[\omega q_1 \sin(\omega t + \theta)\right] \left[V_1 \cos \omega t\right] dt,$$

where the identity $\exp(i\alpha) = \cos\alpha + i \sin\alpha$ has been used. The function $\sin(\omega t + \theta)$ becomes managable by replacing it with the trigonometric identity

$$\sin(\omega t + \theta) = \sin\omega t \cos\theta + \cos\omega t \sin\theta. \tag{7.17}$$

Finishing the calculation (Try it yourself!),

$$\langle P \rangle = \frac{\omega q_1 V_1}{2} \sin\theta, \tag{7.18}$$

or, for the mechanical case,

$$\langle P \rangle = \frac{\omega c_1 F_1}{2} \sin\theta. \tag{7.19}$$

Because phase lies between 0° to −180°, the $\sin\theta$ term is negative, and the power is therefore negative. The minus sign means that mechanical energy is lost; in this LRC circuit through heating of the resistor.

Thus we have learned that the power dissipated by an electrical oscillator is dependent not only on the frequency, charge, and voltage (frequency, amplitude, and force, respectively, for a mechanical oscillator), but also on the sine of the phase angle. The sine of the phase is greatest at kinetic resonance, where the system has maximum current (ωq_1) or speed (ωc_1). While the inductor (mass) and capacitor (spring) conserve energy in this circuit, the resistor (damping) dissipates energy, proportionally to current and voltage (speed and force).

7.1.5 Perturbed harmonic oscillator

Up until now, our model cantilever has been hanging freely, with no tip-sample forces acting upon it other than gravity, which just adds a slight offset to the zero position of the cantilever. Thus gravity is ignored in what follows. If the tip is brought near a surface, van der Waals or other interactions can perturb its motion. We now consider the case of the perturbed harmonic oscillator (PHO), where the behavior of the cantilever is changed only a little. Force control is treated here, which is also a reasonable approximation for displacement control in air or vacuum (Sect. 7.1.1). Excitation and response are assumed to be sinusoidal. This set of conditions applies well to noncontact mode, which was defined and discussed in Sect. 3.4. An increase in the applied load is assumed to result in a proportional increase in the cantilever's amplitude, so the PHO still fits the classification of a linear oscillator.

The second derivative of the tip-sample interaction potential was defined in Sect. 5.1 as k_i, the interaction or interface stiffness, and we placed no restrictions upon its sign or magnitude with respect to k_c. For the purposes of this section, the perturbed harmonic oscillator, we must restrict its magnitude to being significantly smaller than k_c for the oscillator's behavior to be changed only slightly. As a reminder of this restriction, another notation is invoked, Δk, called the perturbation stiffness. If damping is incurred by the interaction, its change is denoted Δb, the perturbation damping. It also must be much smaller than b_c. Both positive (repulsive) and negative (attractive) values are allowed for Δk, but in keeping with physical reality, only positive ones for Δb. The perturbations are accounted for by substituting $(k_c + \Delta k)$ for k_c and $(b_c + \Delta b)$ for b_c in the amplitude and phase equations for the force-controlled SHO, Eqs. 7.7. We then have

$$c_1 = \frac{F_1}{\sqrt{\left(k_c + \Delta k - \omega^2 m\right)^2 + \left(\omega\left(b_c + \Delta b\right)\right)^2}};$$

$$\tan\theta = \frac{-\omega\left(b_c + \Delta b\right)}{\left(k_c + \Delta k - \omega^2 m\right)}. \tag{7.20}$$

Kinetic resonance occurs at a frequency where the $\left(k_c + \Delta k - \omega^2 m\right)$ term goes to zero, whereby the phase angle equals $-90°$, and the amplitude is near its maximum. If $\Delta k = 0$, then kinetic resonance occurs at $\omega_k = \sqrt{k_c/m}$, but if $\Delta k \neq 0$, then the kinetic resonance frequency can shift either up or down, as depicted in Fig. 7.7. A negative (attractive) perturbation stiffness results in a shift downwards, a positive (repulsive) one, upwards. If you were to set your cantilever's drive frequency right at the amplitude maximum, the cantilever amplitude would decrease regardless of the sign of the perturbation stiffness. If the drive frequency were slightly below the free resonance frequency, the amplitude would increase if the peak shifts down in frequency, and decrease if the peak shifts up in frequency. The peak height also changes, because when the $\left(k_c + \Delta k - \omega^2 m\right)$ term equals zero, the frequency

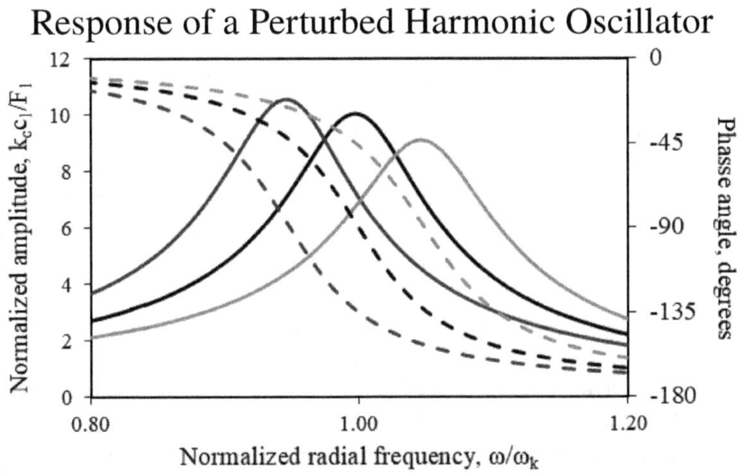

Fig. 7.7 Normalized amplitude (solid lines) and phase angle (dashed lines) as a function of normalized radial frequency for a unperturbed harmonic oscillator (black) with $Q = 10$, the same oscillator perturbed by an attractive interaction (dark gray), and again the same oscillator perturbed by a repulsive interaction with additional damping (light gray).

at which that happens is different from the unperturbed oscillator, such that the remaining $\omega\,(b_c + \Delta b)$ term has a different value, even if $\Delta b = 0$. Additional damping acts to lower the height of the peak.

The changes in the phase angle are correlated with the changes in the amplitude response. Succinctly put, the perturbation stiffness shifts the frequency dependence, and the perturbation damping broadens the frequency range over which the phase changes from in phase (0°) to out of phase (−180°). Because the phase angle changes rapidly near resonance, the cantilever's phase angle is sometimes used in an AFM's feedback system instead of the cantilever's amplitude while using a dynamic mode—a "phase-locked loop." (Sect. 3.4).

★ In **Computer Lab 8**, you examine the behavior of linear and perturbed oscillators on the complex plane. ★

7.2 The nonlinear oscillator

If the forces acting on the tip of the cantilever become strong enough, the behavior of the cantilever can become nonlinear. Intermittent-contact mode AFM (IC-AFM, Sect. 3.4) is usually nonlinear, and the remarks herein are specifically directed to this form of data acquisition. IC-AFM will be explained in the language of oscillators, building upon the ideas in the previous section.

7.2.1 Nonlinear behavior

The adjective nonlinear means that the relationship between a change in the cantilever's environment and its response is not describable by an equation of the form $response = constant \times change$. Instead, there are power laws or exponents in the equation. Nonlinear behavior is intuitively hard to understand and predict. One often observes sudden, seemingly inexplicable, changes in the cantilever motion. Phase contrast can invert, height measurements might not be as expected, and it is easy to misinterpret your data.

Nonlinearity in IC-AFM occurs when nonlinear forces acting on the tip become strong enough such that the interaction stiffness becomes a significant fraction of the cantilever's spring constant k_c. Most forces, and hence interaction stiffnesses, in AFM are nonlinear. Because a nonlinear force can often be approximated as linear over a limited range, the nonlinearity is of little consequence if amplitudes are kept small, as is sometimes true for noncontact mode. The large amplitudes typical of IC-AFM demand consideration of the role of nonlinear forces.

Nonlinearity is more striking near the resonance frequency because of how resonance amplifies the cantilever's movement. The phase of the cantilever response undergoes a 180° change near resonance, and therefore is a sensitive indicator of changes in the forces acting on the tip; thus the wonderful contrast in IC-AFM phase images. Operating near resonance, however, has the drawback that data acquisition parameters also strongly influence the collected image. The higher the quality factor Q, the more of an influence the tip-sample forces have, but high Qs also require the AFM operator to precisely control the data aquisition parameters because of the increased response.

When the strength of the nonlinear forces become great enough, one often observes instabilities in intermittent-contact mode images and force curves. At first, these jumps in amplitude and phase are bewildering. This section attempts to give you a physical understanding of their origin through the derivation and discussion of amplitude spectra. We focus on the instabilities in dynamic force curves, in which the ac amplitude and phase of the cantilever are plotted as a function of tip-sample separation.

7.2.2 Amplitude spectra

This mathematically challenging section should help you develop a better understanding for the origin and behavior of instabilities in IC-AFM data as observed through dynamic force curves. In Sect. 5.1, we explained instabilities in quasistatic force curves in terms of potentials and "balls rolling into valleys." This pleasingly simple imagery, sadly, does not adequately describe the dynamic case. We proceed by using a succession of approximations.

7.2.2.1 Finding the equation of motion

However, our starting point begins once again with familiar ideas of potentials and forces, with which we initially seek the equation of motion of a cantilever subjected to first attractive and then repulsive forces as it is driven farther and farther from its equilibrium position. This time the classic spring potential is perturbed by both an attractive and a repulsive potential,

$$U(z) = \frac{1}{2}k_c z^2 - \frac{1}{4}\xi z^4 + \frac{1}{6}\zeta z^6, \tag{7.21}$$

where ξ and ζ are small with respect to k_c, and the equilibrium position lies at $z = 0$. The attractive (negative) potential goes as the fourth power of z and the repulsive (positive) potential as the sixth power of z. The choice of powers means that the attractive potential has its largest effect when the cantilever is somewhat displaced from its equilibrium position, and the repulsive potential dominates at still greater deflections, much like a cantilever passing through an attractive interaction on its way to tap a sample surface. Figure 7.8a shows both a regular spring potential and the potential of Eq. 7.22. The force is therefore

$$F(z) = -\frac{dU}{dz} = -k_c z + \xi z^3 - \zeta z^5, \tag{7.22}$$

which is drawn in Fig. 7.8b. The cantilever is subjected to this force and driven with an applied load F. The equation of motion of the tip is

$$m\ddot{c} + k_c c - \xi c^3 + \zeta c^5 = F, \tag{7.23}$$

and F is once again an oscillatory function of time, $F_1 \cos \omega t$. This time, in order to keep the calculation simple, we assume no damping and consider only the real part of F (the cosine term).

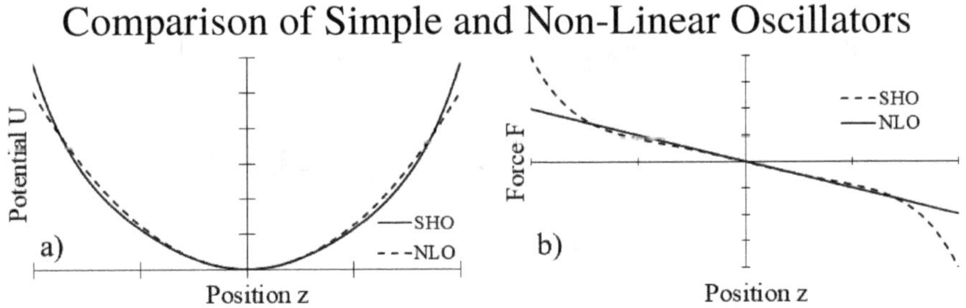

Fig. 7.8 (a) Potentials and (b) forces for a simple harmonic oscillator (SHO, solid lines), such as a spring, and a nonlinear oscillator (NLO, dashed lines), such as a cantilever subject to strong tip-sample forces. Specifically, a parabolic potential and a linear force are used for the SHO, and for the NLO, Eqs. 7.21 and 7.22 are shown in (a) and (b), respectively.

7.2.2.2 Trial solutions

One approach to solving the above equation of motion would be use complex numbers as in Sect. 7.1.1, but it turns out to be easier if we first assume a solution of the form

$$c(t) = c_1 \cos \omega t, \tag{7.24}$$

which we do not expect to work well, yet it might indicate a path to a better answer. Substituting it into the equation of motion and taking the time derivative, one obtains

$$\left(k_c - m\omega^2\right) c_1 \cos \omega t - \xi \left(c_1 \cos \omega t\right)^3 + \zeta \left(c_1 \cos \omega t\right)^5 = F_1 \cos \omega t. \tag{7.25}$$

The intimidating third and fifth powers of the cosine are rendered harmless by the identities

$$\begin{aligned} \cos^3 \omega t &= \frac{1}{4}\left(3 \cos \omega t + \cos 3\omega t\right), \\ \cos^5 \omega t &= \frac{1}{16}\left(10 \cos \omega t + 5 \cos 3\omega t + \cos 5\omega t\right). \end{aligned} \tag{7.26}$$

The right-hand sides of these equations contain what are called odd harmonics, that is, odd numbers times ωt, with the consequence that some of the cantilever motion will be at odd multiples of the drive frequency. Thereby we confirm what we had surmised—Eq. 7.24, with just the drive frequency, is not a very descriptive solution. Based on the trigonometric identities, however, we can try a more sophisticated solution of the form

$$c(t) = c_1 \cos \omega t + c_3 \cos 3\omega t + c_5 \cos 5\omega t. \tag{7.27}$$

The amplitudes c_1, c_3, and c_5 respectively mean the amplitudes of the first, third, and fifth harmonics. Putting this trial solution into the equation of motion, taking time derivatives, keeping only the c_1 terms upon cubing and "quinting" the cosines associated with ξ and ζ (the others are assumed to be small), and lastly rearranging, the next step becomes

$$\begin{aligned} 0 = &\left[\left(k_c - m\omega^2\right) c_1 - F_1 - \frac{3\xi}{4} c_1^3 + \frac{5\zeta}{8} c_1^5\right] \cos \omega t \\ &+ \left[\left(k_c - 9m\omega^2\right) c_3 - \frac{\xi}{4} c_1^3 + \frac{5\zeta}{16} c_1^5\right] \cos 3\omega t \\ &+ \left[\left(k_c - 25m\omega^2\right) c_5 + \frac{\zeta}{16} c_1^5\right] \cos 5\omega t. \end{aligned} \tag{7.28}$$

Each of the terms in square brackets must equal zero, which leads to expressions for c_3 and c_5 as a function of c_1.

$$c_3 = \left(\frac{\xi}{4} - \frac{5\zeta}{16}c_1^2\right)\frac{c_1^3}{\left(k_c - 9m\omega^2\right)},$$
$$c_5 = -\frac{\zeta}{16}\frac{c_1^5}{\left(k_c - 25m\omega^2\right)}. \tag{7.29}$$

7.2.2.3 Constructing an amplitude spectrum

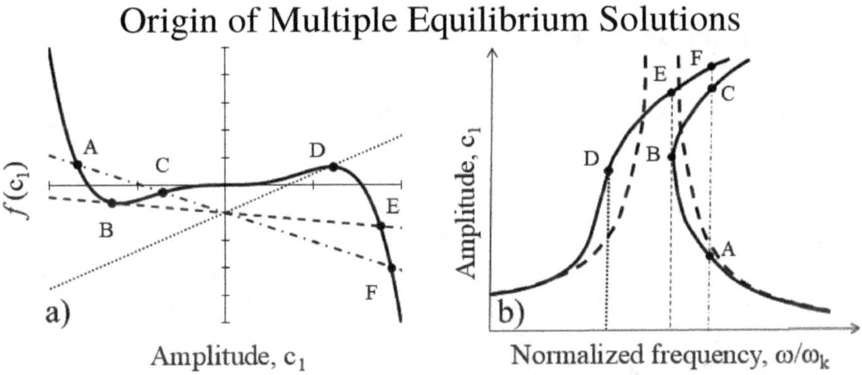

Fig. 7.9 (a) The expressions on the left- and right-hand sides of Eq. 7.30 plotted as a function of $c_1, f(c_1)$. Their intersections (points A, B, C, D, E, and F) correspond to values of c_1 that satisfy Eq. 7.30 for selected normalized frequencies, ω/ω_k. (b) Spectra of the resonances of a simple harmonic oscillator (thick dashed line) and a nonlinear oscillator (thick solid line), both with no damping. The nonlinear spectrum has been constructed from (a) and also displays points A–F. In both (a) and (b), the thin dotted line is for $\omega/\omega_k = 0.85$, the thin dashed line for $\omega/\omega_k = 1.03$, and the thin dot-dashed line for $\omega/\omega_k = 1.10$.

But what about c_1? We need it in order to completely express the solution (Eq. 7.27) in terms of known values. The contents of the first square bracket of Eq. 7.28 can be written as

$$\left[1 - \left(\frac{\omega}{\omega_k}\right)^2\right]k_c c_1 - F_1 = \frac{3\xi}{4}c_1^3 - \frac{5\zeta}{8}c_1^5, \tag{7.30}$$

where the left-hand side is a linear function of c_1 and the right-hand side goes as the cube and fifth power of c_1. We employ a graphical means of finding values of c_1 depending on the normalized frequency ω/ω_k. Figure 7.9a demonstates that if the left- and right-hand sides of the preceding equation are plotted as a function of c_1 for a variety of values of ω/ω_k, the intersections of the curves are the c_1 amplitudes for the corresponding normalized frequency. These pairings of c_1 and ω/ω_k form Fig. 7.9b (solid line). For comparison, the dashed line represents the spectrum of a simple harmonic oscillator. The consequence of ignoring damping is that the peak height is infinite.

There are frequencies for which only one value of amplitude exists, e.g. the frequency corresponding to point D, and there are other frequencies for which three values of amplitude exist, e.g. points A, C, and F. Only one lone frequency is double-valued, that of points B and E. The practical consequence of double- and triple-values is that the amplitude can suddenly change if the conditions are such that more than one value of amplitude is possible. As conditions evolve, a given drive frequency can change from being triple-valued to double-valued to single-valued.

7.2.3 Dynamic force curves

The shape of the nonlinear amplitude spectrum in Fig. 7.9b is significantly different than either the simple harmonic or the perturbed harmonic oscillator, and the existence of more than one amplitude for a given frequency allows the system to jump from one amplitude to another, giving rise to instabilities. Even if you were uncomfortable with the mathematics above, make sure to look carefully at Fig. 7.9b.

We can allow our didactic potential to evolve in a way that idealizes a cantilever-sample approach and retraction—the process of a dynamic force curve—if we first set ξ and ζ equal to zero. Then we allow ξ to increase to some limit. This represents the tip coming into attractive forces. Subsequently, ζ is increased, as we envision the cantilever tapping the sample surface and feeling a repulsive force. The process is reversed to emulate the retraction of the cantilever. The consequent evolution of the amplitude spectra is sketched in Fig. 7.10. The reader's tolerance is requested—while damping was ignored in the calculation, thereby incurring infinite amplitude at resonance, we are now cavalierly assigning finite values to the resonance peak. The vertical dotted lines are to help you keep track of the marks on each spectrum. Dots to the left of the line represent the amplitudes for a below-resonance drive frequency, at the line (squares) the drive frequency is at amplitude resonance, and to the right of the line (diamonds) the drive frequency is higher than resonance. The arrows represent frequencies for which there are two values of amplitude that are about to become single-valued. The arrows show in which direction the amplitude will change. The phase jumps simultaneously, although for clarity, it is not shown here.

The peak first bends to lower frequencies under the influence of the attractive potential, similar to how attraction shifted the perturbed harmonic oscillator's peak in the same direction (Sect. 7.1.5). Then the upper reaches of the peak bend towards higher frequencies, and it continues to fold over towards the right. The right-hand side

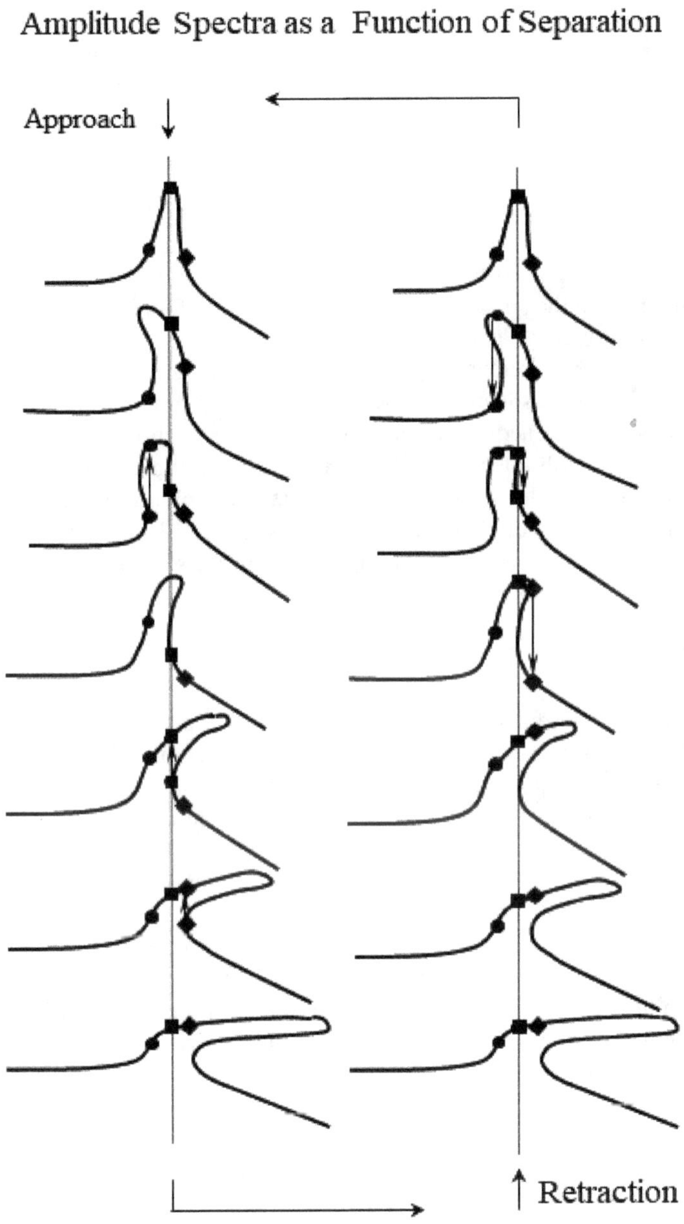

Fig. 7.10 How bistable amplitude spectra evolve during dynamic force curve acquisition. The vertical dotted lines denote amplitude resonance. The pairs of spectra on the left- and right-hand sides are identical. However, the dots, squares, and triangles, which highlight below, at, and above resonance drive frequencies, respectively, are sometimes in different positions for approach and retraction. The instabilities are shown by arrows on the spectra.

of the figure, symbolizing retraction, has the same spectra as on the left (approach). In contrast, the arrows showing the sudden jumps in amplitude are at different positions.

Without question, the presence of instabilities indicates strong nonlinearity in the system. On the other hand, the absence of instabilities does not necessarily mean that the system is linear. The terminology used to distinguish between a system with conditions ripe for instabilities and one that is stable is bistable for the former and monostable for the latter. A bistable system has three solutions to the equation of motion. One is unstable and the others stable. The unstable amplitudes in Fig. 7.10 correspond to those on the underside of the bent-over peak (those that are intermediate to the smallest and greatest amplitude for a given frequency). For a review of stability, read Sect. 5.1.

Figure 7.11 shows sketches of dynamic force curves acquired at drive frequencies above, at, and below resonance for systems that exhibit monostability (left) throughout the range of data acquisition and bistability for portions of the range (right). The cantilever's amplitude is now shown as a function of position rather than of frequency. The jumps in the bistable curves parallel the evolving amplitude in Fig. 7.10. As an exercise, you should label the instabilities in Fig. 7.10 with letters or numbers, then label the corresponding ones in Fig. 7.11 with the same notation.

7.2.3.1 Consequences for imaging

If you were to choose a set of imaging parameters that were bistable, the scanner could abruptly move back and forth between the stable amplitudes appropriate to your imaging conditions, making imaging unreliable. You should, then, choose imaging parameters that lead to monostability. Stability is one of multiple considerations. Others are accurate reproduction of topography, degree of phase contrast, sensitivity to materials properties, and tip-sample pressure (damage and resolution). It is not possible to optimize them all at the same time.

Perhaps the most important choice of imaging parameters is the stiffness of the cantilever. If a cantilever is stiff relative to the highest stiffness of tip-sample interaction that the tip experiences during a cycle of oscillation, then the tip indents the sample for most set-point amplitudes, and the behavior is likely to be completely monostable. In other words, the presence of the sample is not a strong enough perturbation to lead to instabilities. However, by indenting the sample, the "topography" images can be misleading. The tip indents differing materials to different depths. Even with a homogeneous sample, the local sample curvature causes different apparent tip-sample stiffnesses via different reduced radii (Sect. 5.3.4), once again affecting the perceived topography.

In contrast, a weak cantilever relative to the tip-sample interaction is subject to bistability, but has the advantage of giving good topographical data. The sample seems like an infinitely stiff wall to the cantilever. Over the entire sample surface, the tip does not significantly indent the sample, thus reproducing surface contours accurately.

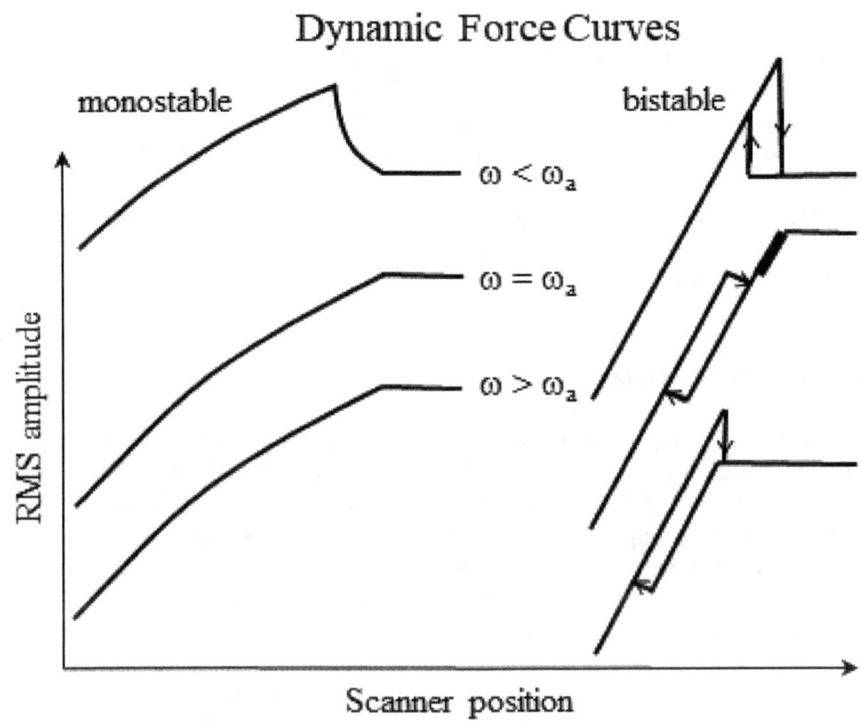

Fig. 7.11 Typical shapes of dynamic force curves. The curves on the left are monostable, where there are no instabilities. The ones on the right are bistable, where there exist two stable solutions for the cantilever's amplitude for some scanner positions. From top to bottom are shown curves for the cases of the drive frequency ω being below, at, and above amplitude resonance, ω_a. The thick line on the right, middle curve shows the best conditions for noncontact dynamic imaging.

As far as phase imaging is concerned, weak cantilevers are so strongly affected by the high repulsive stiffness of the sample that they are not sensitive to the significantly smaller variations in materials properties. The phase is the same for all parts of the sample. Stiff cantilevers are better for phase imaging. Numerical simulations suggest that the deeper the penetration of the tip into the sample, the more the phase is affected by the bulk properties of elastic modulus and damping. Adhesion and local curvature radius become more apparent with less indentation by slightly weaker cantilevers.

Tip-sample pressure and the related issue of resolution can be concerns as well. If the tip or sample are particularly delicate, or if the highest possible lateral resolution is required, one would ideally like the tip to just approach the surface and make the lightest of possible contacts, or perhaps not touch at all. For monostable systems, the cantilever's motion is not significantly modified by attractive forces and the tip and sample touch as soon as the amplitude starts decreasing; the highest workable imaging set point will yield the smallest contact radius and lowest pressures. For bistable systems, the tip and sample are least likely to touch ("true non-contact") at amplitude resonance, with a set-point amplitude above the bistable portion of the dynamic force curve (thicker line in Fig. 7.11). The lower the drive amplitude or the higher the Q of the cantilever, the lower the position of the bistable region on the curve. For drive frequencies lower than resonance, the cantilever leaps into contact at high set-point amplitudes; for frequencies higher than resonance the low-pressure part of the curve is bistable.

Sorting out the best combination of imaging parameters for a particular investigation is greatly aided by the analysis of experimental dynamic force curves and by numerical simulations.

7.2.3.2 Numerical simulations

The fundamental equation in numerical simulations of IC-AFM is the familiar Newton's Second Law, the sum of the forces acting on the tip equal its mass times its acceleration. The additional complication is that the forces are usually nonlinear, and have defined ranges over which they interact. For example, repulsive contact forces do not act on the cantilever if the tip is not in contact. Numerical simulations do not usually add to one's physical understanding of a nonlinear system; hence the derivation of Sect. 7.2.2 and the long discussion above.

However, due to the strikingly nonintuitive behavior of nonlinear systems, numerical simulations are very helpful in predicting how changes in the sample's properties affect the dynamic force curves, and thus give guidance to the experimentalist for choosing image acquisition parameters. Experimental dynamic force curves can also be matched to simulations in order to determine the properties of the sample, but here the note of caution is that the equation of motion can have up to five unknowns. If more than one parameter is unknown, the input parameters that yield the matching simulated data are not unique. Finally, simulations allow study of data that is hard to determine experimentally, for example, the area of tip-sample contact or the tip-sample pressure.

So ends the chapter and book. We hope you have been able to build a solid foundation in AFM and that you will continue your adventures "in force."

7.3 Learning objectives, further reading, and problems

! Learning objectives for Chapter 7

7.1 Define force and displacement control and describe when they differ in behavior.
7.2 Be able to use the equations for amplitude and phase of an SHO and a PHO.
7.3 Sketch the amplitude and phase of an SHO. Be able to label the axes with both unnormalized and normalized variables.

7.4 Sketch how amplitude and phase depend on attractive and repulsive perturbations for a PHO.
7.5 Distinguish between linear and nonlinear systems and between monostability and bistability.
7.6 Describe the complexities of acquiring good images in intermittent-contact mode.
7.7 Know how numerical simulations of dynamic force curves can be helpful or misleading.

Further Reading

✓**History.** Burnham, N. A., Behrend, O. P., Oulevey, F., Gremaud, G., Gallo, P. J., Gourdon, D., ... & Briggs, G. A. D. (1997). How does a tip tap?. Nanotechnology, 8(2), 67.

✓**History.** Cleveland, J. P., Anczykowski, B., Schmid, A. E., & Elings, V. B. (1998). Energy dissipation in tapping-mode atomic force microscopy. Applied Physics Letters, 72(20), 2613–2615.

✓**Book.** Marion, J. B. (2013). Classical dynamics of particles and systems. Academic Press.

✓**Tutorial.** McMullen, C. P. D. (2018). Differential Equations: Essential Skills Practice Workbook with Answers. Zishka Publishing.

? Problems for Chapter 7

7.1 Why isn't the gravitational force shown in Fig. 7.1?

7.2 Provide the steps between Eqs. 7.2 and 7.6.

7.3 Provide the steps between Eqs. 7.6 and 7.7.

7.4 Show how Eqs. 7.8 are obtained from Eqs. 7.7.

7.5 Take the (a) low-frequency, (b) kinetic-resonance, and (c) high-frequency limits for Eqs. 7.7 and 7.8.

7.6 Show how Eqs. 7.10 are obtained from Eqs. 7.9.

7.7 Take the low-frequency, kinetic-resonance, and high-frequency limits for Eqs. 7.9 and 7.10.

7.8 Verify that the four tangent values for the four shown normalized frequencies of Fig. 7.5 are correct.

7.9 Provide the steps between Eqs. 7.12 and 7.13 by using Eqs. 7.5.

7.10 Solve for the equivalent of Eqs. 7.7 for an LRC circuit, by using using Eqs. 7.5.

7.11 Write out the equivalent of Eqs. 7.7 for an LRC circuit, by using the equivalent variables discussed in Sect. 7.1.4.1.

7.12 Provide the steps between Eq. 7.15 and Eq. 7.18.

7.13 Instead of considering the real parts of current and voltage for Eq. 7.15, perform the same integration with the imaginary parts over one period, $\langle P_{Im} \rangle$, and compare it with Eq. 7.18. Does your answer make sense, and if so, why?

7.14 Take (a) the low- and (b) high-frequency limits for Eqs. 7.20. Also (c) find c_1 and $\tan\theta$ when $\omega = \sqrt{(k_c + \Delta k)/m}$.

7.15 Sketch Fig. 7.9b and indicate which ranges of frequencies are single-valued, double-valued, or triple-valued in amplitude.

7.16 Match the amplitude transitions in Fig. 7.10 to those in Fig. 7.11 by making sketches of the curves and labeling each set of transitions with a letter.

7.17 Sketch the set of three bistable dynamic force curves in Fig. 7.11, then shade the set-point amplitudes at which you would expect unreliable imaging.

7.18 Look online to find out what drive frequency and set-point amplitude an AFM manufacturer recommends for dynamic-mode imaging. Your instructor might organize your class such that each student will look for a different manufacturer's recommended protocol. Then the class can discuss the similarities and differences.

7.19 Lab question. Can you replicate the dynamic force curves in Fig. 7.11?

7.20 *Challenge Problem.* Derive Eq. 7.11.

7.21 *Challenge Problem.* Provide the steps between Eqs. 7.23 and 7.28.

Appendix A
Compiled learning objectives

Dear Reader,

Thank you for choosing this book. I hope learning about AFM has been enjoyable for you. Here—for your convenience—I compiled the learning objectives from the book's seven chapters. If you are comfortable with all of them, then you have a good grasp of AFM.

May the Force be with you!

Professor Nancy A Burnham, PhD

A.1 Chapter 1, Context and overview of instrumentation

1.1 Give examples of the size scale of AFM.
1.2 Distinguish between "top-down" and "bottom-up."
1.3 Demonstrate the operational principle of AFM.
1.4 List the disadvantages and advantages of AFM.
1.5 Describe a typical probe's size and composition.
1.6 Understand how an AFM collects $d(x, y)$ and $z(x, y)$ data.
1.7 Relate common AFM-related acronyms to their function.
1.8 Be familiar with basic image-processing options.

A.2 Chapter 2, Difficulties of imaging

2.1 Sketch a diagram of an AFM system.
2.2 Define the error signal.

2.3 Understand the purpose of the feedback circuit.
2.4 Recognize when the feedback gains are too high, about right, and too low.
2.5 Demonstrate image optimization using feedback and other scan parameters.
2.6 Be familiar with common artifacts and how to test and correct for them.
2.7 Describe the common features of a noise spectrum.
2.8 Know how to situate and protect your SPM.
2.9 Understand how Fourier filtering can improve noisy images.
2.10 Acquire and process topographic AFM images.

A.3 Chapter 3, Other SPMs and AFM imaging modes

3.1 Compare and contrast an SPM, an STM, an AFM, and an LFM.
3.2 Discuss STM's experimental requirements.
3.3 Describe by analogy what tunneling current is.
3.4 Know how static, sliding, and average friction are determined from a friction loop and what influences them.
3.5 Given the topography and friction coefficients of a hypothetical surface, sketch the expected friction loop.
3.6 Distinguish among contact, non-contact, intermittent-contact, off-resonance, and lateral-force modes of AFM operation.
3.7 Know the advantages and disadvantages of the different AFM modes of operation and understand when they are used.

A.4 Chapter 4, AFM design and calibration

4.1 Distinguish among the parts of a probe assembly.
4.2 Understand why certain materials and geometries are used to construct probes.
4.3 Discuss the phrases"range versus resolution" and "robustness versus resolution."
4.4 Describe the various ways in which tips and cantilevers are calibrated.
4.5 Describe the various ways in which deflection and scanners are calibrated.
4.6 Be familiar with common problems of scanners.
4.7 Explain how the scanner problems can be corrected.

A.5 Chapter 5, Force-curve mechanics

5.1 Know the basic physical principles behind force curves—the relationships among potentials, forces, and concavity.
5.2 Be able to determine if an equilibrium is stable, unstable, or neutral, both mathematically and by inspecting a potential diagram.

5.3 Define monostability and bistability.
5.4 Explain the origin of cantilever instabilities.
5.5 Define a force curve.
5.6 Understand how raw force-curve data are converted to processed data.
5.7 Discuss the limitations of weak cantilevers for force-curve aquisition.
5.8 List and describe common problems with force curves.
5.9 Compare and contrast elasticity, plasticity, and viscoelasticity.
5.10 Be able to sketch and discuss a stress-strain curve; define σ, ϵ, and E.
5.11 Define υ, E_r, and K.
5.12 Define R_r, a, and ϖ.

A.6 Chapter 6, Tip-sample interactions

6.1 Distinguish between attraction and adhesion.
6.2 Describe the origin of the van der Waals potential.
6.3 Understand how the van der Waals potential takes different functional forms for different geometries.
6.4 Interpret the features of a force curve due to meniscus and capillary formation.
6.5 Sketch how the van der Waals and meniscus-capillary forces manifest themselves in force curves taken with weak and stiff cantilevers.
6.6 Identify the principal features of a force vs depth curve.
6.7 Be able to manipulate and use the Hertz equations.
6.8 Give reasons why there are other contact mechanical models.
6.9 Distinguish between plasticity and adhesion hysteresis.
6.10 Discuss the advantages and disadvantages of the continuum versus molecular dynamics approaches.

A.7 Chapter 7, Oscillator models for AFM

7.1 Define force and displacement control and describe when they differ in behavior.
7.2 Be able to use the equations for amplitude and phase of an SHO and a PHO.
7.3 Sketch the amplitude and phase of an SHO. Be able to label the axes with both unnormalized and normalized variables.
7.4 Sketch how amplitude and phase depend on attractive and repulsive perturbations for a PHO.
7.5 Distinguish between linear and nonlinear systems.
7.6 Describe the complexities of acquiring good images in intermittent-contact mode.
7.7 Know how numerical simulations of dynamic force curves can be helpful or misleading.

Appendix B
General plan of an AFM experiment

How do you go about planning and executing an experiment? Most experiments proceed in the four major steps outlined here.

B.1 Plan

Think about what you want to learn about your sample, in which size and shape you can prepare your sample, and what is the likely magnitude of topography. Is there an AFM available to you that has the appropriate scan and z-range? Similarly, do you need a special environment, such as fluids or vacuum? How are the samples placed into the AFM? How well calibrated do the data need to be? If the tip or sample might be damaged by your experiment, what sample can you use as a standard that you can examine at the beginning and end of your experiment to check for consistency?

Think about which data acquisition mode might be best for learning about your sample. Does the AFM offer this mode? Choose an appropriate probe for your experiment, considering cantilever spring constant and tip size and shape. Purchase the probes, sample mounts, and any calibration materials that are not already available to you.

Review Chap. 3 for data acquisition modes and Chap. 4 for probe design and instrument calibration. Learn about the AFM you will use; the steps below apply to most AFMs, but the one you use might be slightly different. Always follow the instructions given to you by the manufacturer or manager of the AFM.

Also consider complementary techniques, such as optical and scanning electron microscopies (Chap. 1) that might reinforce (or contradict) what you might learn with AFM.

B.2 Begin

- Turn on the hardware (if it is not already on).
- If the tip or sample might be modified by your experiment, mount your standard sample. If modification is less likely, mount your sample of interest. Turn on the software.
- Most programs will ask you to select the probe from its database. They might also ask about the sample and desired acquisition parameters. Choose the data acquisition mode.
- Ensure that the excitation light (laser, in most cases) reaches the cantilever (ideally at its free end) and light is reflected towards the center of the photodiode.
- Check that the probe will land on a relevant part of the sample. Advance the probe towards the sample until about a millimeter separates them, then activate the auto-approach.

B.3 Acquire

- Start imaging. Choose a display filter that allows you to see the relevant aspects of the sample.
- Optimize the scan parameters by minimizing the error signal. Check for noise and artifacts (Chap. 2).
- If relevant, acquire force curves (Chaps. 5 & 6).
- Check if your data acquisition has damaged the sample or modified the tip, possibly returning to your standard sample.
- Save the data in its original form, such that any image filters that you applied during acquisition do not distort the data.

B.4 Process and Present

- Most images require at least background subtraction, if not more advanced image processing options (Chap. 2).
- It is easier for the reader to compare images if the same color scale is used. Or if not the same color scale, color scales in integer multiples of each other will help the reader comprehend more quickly what you have observed.
- Make sure to label your axes with units and use large enough fonts such that readers do not have to strain to see the results.
- Figure captions should help the reader understand what is the important aspect of the figure.

The four main steps—Plan, Begin, Acquire, Process—create a rhythmic acronym: P-BAP.

Appendix C
Electronic supplement

Dear Reader,

In my courses, I try to give students multiple perspectives on the course content, and as such, I have created a variety of materials that complement this textbook. The materials are available free to you after filling out an online form with your contact information and how you plan to use the material, e.g., in a course or for research, using the QR code at the end of this appendix. Short descriptions of each of the available electronic files are listed below.

Yours in Force, Professor Nancy A Burnham, PhD

C.1 Online videos, Canvas and Moodle questions

The online videos summarize the course content. Most of them were created in 2013, when the easily available video processing programs were still new. In time, I will revise the videos to modern standards of appearance and pace. They are organized into four playlists at the AtomicForceMicro channel on YouTube:

1. Tutorials for use of the NaioAFM, example student projects (12 videos, 70 min)
2. Fundamentals of AFM (18 lessons, 3.5 hrs)
3. Interpretation of AFM force curves (12 lessons, 2.6 hrs)
4. Minute monologues on example published research (11 videos, 15 min)

While the YouTube channel is publicly available, multiple-choice quizzes on the videos are provided in the supplement for the learning management systems Canvas (for my courses at WPI) and Moodle (for my course at ETH Zurich). For the longer undergraduate and graduate courses at WPI, there are sixteen content quizzes and six pre-lab quizzes. For the two-week block course at ETH Zurich, there are six content quizzes and six pre-lab quizzes.

C.2 Instrument lab instructions and practical quiz

Instructions for six instrument labs with a Nanosurf NaioAFM are furnished. The NaioAFM was chosen for its reliable laser-cantilever-photodiode alignment—a common hurdle for new users—and its modest expense. Naturally, you would be welcome to adapt the instructions to your AFM. The topics are:

1. Laboratory procedures and basic imaging
2. Scan parameters and display options
3. Optimizing an image
4. Scanner and probe calibration
5. Acquiring and processing force curves
6. Material identification

A practical quiz is also provided. The practical quiz is given at the beginning of the second lab session. The students are told during the first lab session that as a group they will be asked to acquire an image without direct help from the teaching assistant, motivating them to take good notes during the first lab session.

C.3 Computer lab instructions, software, and installation

The computer labs are to help the students learn the basics of image processing and to illustrate the more difficult course concepts. The image processing lab is the first one. It is based on Gwyddion, an open-source program publicly available at *gywddion.net*; the instructions were written for V.2.66. The subsequent computer labs were written by me.

My courses cover the first six or seven of the following list:

1. Image processing
2. Feedback and noise
3. Fast fourier transforms and fourier filtering
4. Potentials, forces, and stiffnesses
5. Surface forces and cantilever stiffness
6. Contact mechanics
7. Molecular dynamics
8. Linear oscillators

C.4 Macroscopic-cantilever lab instructions

Students sometimes have trouble visualizing what the cantilever does, and thus I developed "macro-labs" with a macroscopically sized cantilever. The macro-labs take just a few minutes to perform. The instructions for the six macro-labs cover these topics:

1. Spring constant by two techniques
2. Tip imaging
3. Amplitude and frequency shifts
4. Cantilever instabilities
5. Spring constants of unknown cantilevers and system
6. Reduced modulus from the Hertz equation

C.5 Workbook for force-curve parameters

With cantilevers ranging in stiffness from ~0.1–100 N/m and tip radii ranging from nanometers to microns, it can be difficult to plan a force-curve experiment. The workbook entitled *Hertz DMT BCP calculations.xlsx* can help you decide which probes to buy and which experimental parameters to use. Furthermore, if you use all three included spreadsheets (Hertz, DMT, BCP), you will see how your choice of model affects the most appropriate selection of experimental parameters.

C.6 Templates for daily summaries and report, grading rubric

Most instructors want to see students turn in reports of their lab work. In my two-week block course, I ask the students to write daily summaries. In my longer courses of seven (undergraduate) or fourteen (graduate) weeks, I ask the students write formal lab reports for the instrument labs and informal reports for the computer and macro-labs. The templates for the daily summaries and instrument lab reports are included in the supplement, as well as the grading rubric for the lab reports.

C.7 Worksheets, writing exercises, and evaluation forms for class

In my longer courses, I usually run a "flipped" classroom, in that I ask the students to watch the online videos and answer quiz questions before class. This choice frees up time during class for discussion, problem-solving, writing exercises, and student presentations. The worksheets are problem sets to help the students solidify the material. The writing exercises are to assist them in thinking about scientific writing; my goal is that by the end of the course, they are able to independently draft the equivalent of an Applied Physics Letter. The evaluation forms are what I provide to the students before they submit an abstract and present a talk on a project, be it a literature review or their laboratory experiments.

You may request the complementary materials associated with this book via the QR code below:

Appendix D
List of acronyms

AFM	Atomic force microscope, atomic force microscopy
BCP	Burnham, Colton, Pollock contact mechanics
DMT	Derjaguin-Muller-Toporov contact mechanics
FFM	Friction force microscopy
FFT	Fast Fourier transform
IC-AFM	Intermittent-contact mode AFM
JKRS	Johnson-Kendall-Roberts-Sperling contact mechanics
LFM	Lateral force microscopy
LVDT	Linear variable differential transformer
NC-AFM	Non-contact mode AFM
PHO	Perturbed harmonic oscillator
PI	Proportional, integral feedback
PID	Proportional, integral, differential feedback
PPM	Proximal probe microscopy
PTFE	Polytetrafluoroethylene
SEM	Scanning electron microscope
SHO	Simple harmonic oscillator
SPM	Scanning probe microscopy
STM	Scanning tunneling microscopy
SXM	Scanning "X" microscopy

Appendix E
List of symbols

$0,\ldots,n$	Subscripts: 0 refers to dc, 1 to ac, n to higher harmonics
c, i, tot	Subscripts: cantilever, interaction, total, respectively
m, r	Subscripts: medium and reduced, respectively
s, t	Subscripts: sample and tip, respectively
a	contact radius
a_i	charge separation within a dipole
b	damping coefficient
c	position of backside of cantilever above tip
d	deflection of cantilever, $c - h$
f	an arbitrary function
h	position of cantilever holder or head of microscope
h	In Eq. 6.11, Planck's constant, 6.626×10^{-34} J·s
$\hbar$	Planck's constant divided by 2π, 1.055×10^{-34} J·s
i	$\sqrt{-1}$
k	spring constant, stiffness, concavity
k_B	Boltzmann's constant, 1.38×10^{-23} J/K
l	length
m	effective mass of cantilever
m_e	mass of electron, 9.11×10^{-31} kg = 0.511 MeV/c^2
m_{vol}	volumetric mass of cantilever
n	index of refraction
o	number of oscillations
q	charge
r	interatomic distance
$\mathbf{r}$	position vector
$\hat{r}$	unit vector in outward radial direction
s	position of sample surface
t	time
v	volume
ϑ	velocity

w	width
x	horizontal coordinate axis
$\hat{x}$	unit vector in $+x$ direction
y	vertical coordinate axis
$\hat{y}$	unit vector in $+y$ direction
z	coordinate axis normal to sample surface
$\hat{z}$	unit vector in $+z$ direction
z_0	constant position along z-axis
A	Area
B	Proportionality constant for an interatomic potential
C	Capacitance
E	Young's or elastic modulus
$\mathbf{E}$	Vector Electric field
E_{tot}	Total energy
F	Magnitude of force on tip or sample
$\mathbf{F}$	Vector force on tip or sample
H	Hamaker constant
I	Current
J	Compliance
K	Indentation modulus
K_E	Kinetic energy
L	Inductance
N	Normal force from sample on tip
P	Power
P_H	Hertzian pressure
Q	Quality factor
R	Resistance
T	Time interval
U	Potential
V	Voltage
W	Work
Y	Admittance
α	a positive constant
β	a positive constant
δ	separation or indentation depth, $c - s$
γ	surface energy (solids) or surface tension (liquids)
ϵ	strain
ε	dielectric constant
ε_0	permittivity of free space
ζ	a small constant
η	tip height
θ	phase difference between response and excitation
$\hat{\theta}$	unit vector in the $+\theta$ (ccw) direction

κ tunneling wavenumber
μ coefficient of friction
ν frequency in Hertz
ξ a small constant
ρ volume density
ϱ number density
σ stress
ς viscosity
τ thickness of cantilever
υ poisson ratio
χ polarizability
ω radial frequency
ω_a amplitude resonance frequency
ω_k kinetic resonance frequency
ϖ work of adhesion

Δ Change in a variable
Θ Angle of repose of cantilever
Υ Temperature
Φ work function

Answers

Chapter 1 Answers

1.1 This should be easy!

1.2 (a) 1.61×10^9 μm; (b) 1.61×10^9 nm; (c) 1.61×10^9 pm

1.3 1.8 nN

1.4 260 Pa

1.5 180 MPa

1.6 300 GPa

1.7 (a) 250 MPa = yield strength of A36 steel; (b) No. Damage is expected.

1.8 See Fig. 2.1.

1.9 Constant-height mode: no feedback, scanner(s) remain at same z-position; Constant-force mode: with feedback, scanner(s) move in z to try to keep force constant while also rastering in x and y.

1.10 Constant-height mode is faster because the feedback is not engaged. In constant-force mode, extremes of force are nominally avoided during imaging, for example, losing contact with the sample or incurring high enough force to cause damage to the tip or sample.

1.11 See Sect. 1.1.2.

1.12 Both AFM and SEM make images while rastering the excitation relative to the sample. Both can cause damage to the sample.

1.13 **Lab question.** Your instructor will tell you the expected feature spacing and heights of the sample that was provided to you for your first AFM images. (Congratulations!)

Chapter 2 Answers

2.1 (a) 330 nm; (b) 1.1 V

2.2 32 ng

2.3 Areas 1&2, too high gain; Area 3, about right gain; Area 4, too low gain

2.4 (a) 195 nm; (b) 100 μm/s; (c) 2.0 ms

2.5 1.3 kHz

2.6 While not every online AFM image suffers from artifact(s), a surprising proportion do. That's why we offer this course!

2.7 (a) 0; (b) $Y^2/3$; (c) $Y\sqrt{3}$

2.8 The only non-zero fourier coefficient is $a_1 = 1/2$.

2.9 **Lab question.** Raw data vs filtered data scanning along steps, since each scan line is set to an average value.

2.10 **Lab question.** From greatest to least influence: I, P, D

2.11 **Lab question.** See Sect. 2.3.

2.12 *Challenge Problem.* Hint: first square $f(t)$, then integrate, looking for functions that sum to zero over an interval of one period.

Chapter 3 Answers

3.1 (a) atomic forces; (b) tunneling current; (c) lateral atomic forces; (d) for example, magnetic and electric forces, ionic current, reflected light, etc.

3.2 This should be easy!

3.3 This too!

3.4 For both (a) and (b), both sides of the equation are in inverse distance.

3.5 (a) 9.58 1/nm; (b) Yes; (c) 1.5 μA, 4.8×10^{-14} A.

3.6 (a) 11.5 1/nm; (b) 1.0 μA, 0.10 μA; (c) 3, 2; (d) Because the input variables have that many "sig figs," respectively.

3.7 (a) Polycarbonate, because in the LFM images, the difference in color scale between the forward and backward images is greater than for the aluminum. (b) Yes. (c) The LFM images directly record the twisting of the cantilever, so you see the start-up artifacts as narrow bands on the left- and right-hand sides of the forward and reverse LFM images, respectively. The topographic images are less sensitive to the start-up artifact caused by friction in contact mode because they rely on the up-and-down motion of the cantilever.

3.8 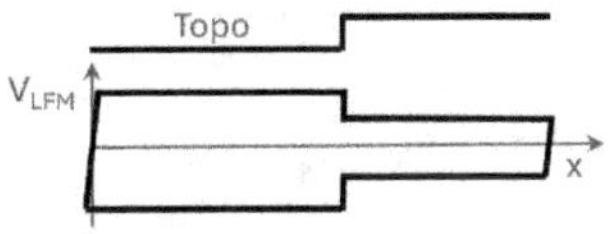

Note that the nearly vertical sections of the friction loop are indeed not vertical, accounting for the small start-up artifacts in the lateral-force image. In the next problem, we address some details regarding the twisting as the tip goes up and down steps. Here, the one step along this fast-scan line does not cause significant spikes.

3.9 (a) Spikes 1 and 4 represent static friction being higher than sliding friction; at the start of a scan, the tip is momentarily at rest. Spikes 2 and 5, pointing away from the x-axis, are when the tip twists a lot as it goes up the step. Spikes 3 and 5, pointing towards the x-axis, are when the tip stands up straighter as it comes down the step. (b) Yes, correct. There is hysteresis between forward and backward LFM data, corresponding to the motion of the scanner during the twist from one orientation to the other as the raster scan changes direction. To a lesser extent, the cantilever itself also bends laterally; see Sect. 4.1.1.

3.10 Check your sketches with your instructor.

3.11 (a) $1 = \mu N \sin\theta$; $2 = \mu N \cos\theta$; $3 = N \sin\theta$; $4 = N\cos\theta$. (b) F_{lat}, μN, 1, 2 all reverse their directions. F_{load} and N and their components remain the same. (c) $\sum F_x = 0 = -F_{lat-} - N\sin\theta + \mu N\cos\theta, \sum F_z = 0 = -F_{load} + N\cos\theta + \mu N\sin\theta$.

3.12 (a) $c_1 = F_1/\sqrt{k_c^2 + (\omega b_c)^2}$, $\tan\theta = 0$; (b) $c_1 = F_1/\omega b_c, \tan\theta = -\infty$; (c) $c_1 = 0$, $\tan\theta = 0$.

3.13 (a) Unitless; (b) kg/s; (c) If c_1 is in m and F_1 in Newtons, then the denominator on the RHS must be in N/m. It is, for k_c is squared, then squared-rooted, and the other terms under the square root also have units equivalent to N/m, as you should show.

3.14 (a) 2.00×10^{-13} J = 1.25 MeV; (b) 18.6 cm/s; (c) 100 nm.

3.15 Confer with your instructor.

3.16 Delayed by a compliant viscous sample with high adhesion (or any of these three factors). Advanced by a stiff elastic sample with low adhesion (or any of these three factors).

3.17 (a) Solve Eq. 3.3b for N and substitute it into Eq. 3.3a to arrive at Eq. 3.4a. Do the same for your answers to P. 3.11c. (b) Equation 3.6a is the average and Eq. 3.6b is the difference of of Eq. 3.5.

Chapter 4 Answers

4.1 (a) 1/8; (b) 1/8; (c) 1/2; (d) 1/4

4.2 (a) 8; (b) 2; (c) 8; (d) 2

4.3 (a) 225; (b) 400

4.4 (a) $33\rho w\tau l/140$; (b) $\rho w\tau l$; (c) See Sects. 2.1 and 4.1.1.2.

4.5 (a) Short, wide, thick, short tip; (b) $(G/6\pi)\sqrt{35E/(33\rho)}w\tau^4/(\eta^2 L^3)$; (c) No

4.6 (a) Short, narrow thin; (b) $(\pi/2)\sqrt{33\rho E/35}(w\tau^2/l)$; (c) No

4.7 (a) Long, wide, thin, short tip; (b) $4\varsigma l^4 w^2/(E\tau^3\eta^3)$; (c) No

4.8 Answer depends on customer's needs.

4.9 Your choice of probe should be consistent with your customer's needs.

4.10 (a) 636 pm, 6.36 pm; (b) 74.3 pm, 743 fm

4.11 (a) Eq. 4.2 is in N/m; (b) Eq. 4.5 is in 1/s; (c) Eq. 4.8 is in N/m.

4.12 (a) Left to right; (b) The steeper and shallower slopes reverse sides.

4.13 Confer with your instructor.

4.14 (a) 100 Ω; (b) 101 Ω.

4.15 (a) Both sides of the equation are in Coulombs per Volt. (b) 88.5 pF; (c) Sketch goes as $1/d$, which is nonlinear.

4.16 (a) Both sides are in T = N/A·m. (b) 126 μT.

4.17 **Lab question.** Many societal nightmares could ensue if scientific conclusions were based on poor calibration of the tip, cantilever, or scanner.

4.18 *Challenge Problem.* (a) $(2\pi\tau\nu)^{3/2}(w/4)(33\rho/35E^3)^{3/4}$; (b) N/m

4.19 *Challenge Problem.* (a) $\Delta z(1 + \cos 2\Theta)$; (b) $\lambda/(1 + \cos 2\Theta)$.

Chapter 5 Answers

5.1 (a) a-one, b-one (curves 3,4,8) or three (curves 5,6,7); (b) Curve 5 in Fig. 5.4b looks to be neutral (flat); (c) a-all are monostable, b-curves 3,4,8 are monostable, curves 5,6,7 are bistable.

5.2 Approach-between curves 4 and 3, Retraction-between curves 7 and 8

5.3 $F = -kx$ and k.

5.4 Equilibria at $x = 0$ (stable), $-2/3$ (unstable)

5.5 (a) $-k_c(z-h) + (4\alpha/\beta)[(\beta/z)^5 - (\beta/z)^3]$; (b) $-k_c + (4\alpha/\beta^2)[5(\beta/z)^6 - 3(\beta/z)^4]$

5.6 (a) $1/k_r = 1/k_c + 1/k_i$ or $k_r = k_c k_i/(k_c + k_i)$; (b) $k_r^{||} = k_c + k_i$

5.7 The denominator becomes close to k_i, and the k_i's cancel.

5.8 On the z-axis, $c > s > h$, the directions of vectors **d** and $\boldsymbol{\delta}$ flip, as do the forces of the cantilever and interaction on the tip.

5.9 (a) $\Delta\delta = 0$; the slopes will be infinite. (b) $\Delta\delta = \Delta d(Error)$; the $d(\delta)$ curve's slope equals $Error$; the $F_i(\delta)$'s slope equals $k_c(Error)$. (c) $\Delta\delta = \Delta d$; the $d(\delta)$ curve's slope equals one and the $F_i(\delta)$ curve's slope equals k_c.

5.10 (a) $-1 + (\eta/2l)(\tan\theta \pm \mu)$; (b) Unloading.

5.11 Pythagorean theorem for closest approach, drop small terms, definition of $\tan\theta$.

5.12 (a) $x_0 y_0 z_0(1 + \epsilon_x + \epsilon_y + \epsilon_z)$; (b) $(1 + \epsilon_x + \epsilon_y + \epsilon_z)$; (c) $1 - 2\upsilon\epsilon_z$; (d) 0.5

5.13 (a) $(1 - \upsilon_s^2)/E_s$; (b) $2(1 - \upsilon^2)/E$; (c) $(1 - \upsilon_t^2)/E_t$

5.14 (a) $\delta/(8R)$; (b) Yes; (c) No

5.15 $dz/dx = \tan\theta \rightarrow \epsilon_{max} = \delta/(2R - \delta)$.

5.16 (a) $R_t = R_s = R \rightarrow R/2$; (b) $R_s = \infty \rightarrow R$; (c) $R_s = -2R \rightarrow 2R$

5.17 The hypotenuse of the triangle is R; its sides are a and $R - \delta/2$.

5.18 After using Eq. 5.16, cancel the a's.

5.19 Generally, researchers interested in adhesion present the data as a function of scanner position, and researchers interested in elastic modulus present the data as a function of indentation depth.

5.20 **Lab question.** Confer with your lab instructor.

5.21 *Challenge Problem.* Figure 5.3 implies that every step in z compresses or extends the springs by the same amount, yielding parallel behavior, $k_r^{||}$. In contrast, in a true series configuration, the springs will compress or expand depending on their relative strength, ending with k_r.

5.22 *Challenge Problem.*
(a) $F_i(z) = -dU_i/dz = -(4\alpha/\beta)[(\beta/z)^5 - (\beta/z)^3]$, $k_i(z) = -dF_i/dz = (4\alpha/\beta)[5(\beta/z)^6) - 3(\beta/z)^4]$;
(b) Equilibrium is at $dF_i/dz = 0$, so $z_0 = \beta$;
(c) $U_i(\beta) = -\alpha$, $k_i(\beta) = 8\alpha/\beta$;
(d) $\sqrt{5/3}\beta$;
(e) -0.84α, $-0.72\alpha/\beta$;
(f) The sketch should look similar to U_i in Fig. 5.4.
(g) The equilibrium position is β, and the potential at equilibrium is $-\alpha$.

Chapter 6 Answers

6.1 (a) Force vectors point away from the charges that create them. (b) As in (a), by Newton's Third Law. (c) Net force vectors point towards each other.

6.2 (a) Volume integration by $\varrho_s dv = 2\pi \varrho_s x dx dz$; (b) Volume integration by $\varrho_f dv = A dz$, with $H = \pi^2 \varrho_f \varrho_s B$; (c) Volume integration by $2\pi \varrho_t R(z - \delta)\, dz$

6.3 $F_{t-s} = -HR/(6\delta^2)$

6.4 (a) Air: $H_{\omega=0} = 0.864$ meV, $H_{\omega>0} = 0.753$ meV; (b) Acetal: $H_{\omega=0} = 0$, $H_{\omega>0} = 0.562$ meV; (c) Air: the two contributions are about the same. Acetal: only the fluctuating dipoles contribute.

6.5 (a) Positive. Numerators and denominators of $H_{\omega=0}$ and $H_{\omega>0}$ are positive. (b) $H > 0$ because the numerators and denominators are again all positive. (c) When the dielectric constant and index of refraction of the medium are in between that of the tip and sample, then the two contributions to H are both negative (numerators negative, denominators positive). (d) Choose tip and sample to be the same material with no permanent dipoles and match the indices of refraction with the medium.

6.6 (a) $-2\pi\gamma r_l$; (b) -4.52 nN

6.7 In N, N/m, and N/m^2, respectively.

6.8 (a) 316 nN; (b) 0.474 N/m; (c) 158 nm; (d) 1.16 μm.

6.9 (a) 1000 nN; (b) 150 N/m; (c) 5.0 nm; (d) 15 nm.

6.10 (a) $dF/da = 3Ka^2/R$,
$dF/d|\delta| = 3K(R|\delta|)^{1/2}/2$,
$dk_i/da = 3K/2$,
$dk_i/d|\delta| = 3KR^{1/2}/4|\delta|^{1/2}$,
$dP_H/da = K/\pi R$,
$dP_H/d|\delta| = KR^{1/2}/2\pi|\delta|$;
(b) Constant: dP_H/da, dk_i/da;
(c) Infinity: $dk_i/d|\delta|$, $dP_H/d|\delta|$.

6.11 N for force and m for depth.

6.12 (a) $\bar{F} = \bar{A}^3$, $\bar{\Delta} = -\bar{A}^2$; (b) $\bar{F} = \bar{A}^3 - 2$, $\bar{\Delta} = -\bar{A}^2$; (c) $\bar{F} = \bar{A}^3 - \sqrt{3\bar{A}^3/2} - 1$, $\bar{\Delta} = 1 - \bar{A}^2$; (d) $\bar{F} = \bar{A}^3 - \sqrt{6\bar{A}^3}$, $\bar{\Delta} = (2/3)\sqrt{6\bar{A}} - \bar{A}^2$.

6.13 (a) $P = Ka/\pi R - \sqrt{3\varpi K/2\pi a} - \varpi R/a^2$; (b) $dF/da = 3Ka/R - \sqrt{27\pi\varpi Ka^2/8}$; (c) The pressure is large and negative, proportional to $-1/a^2$, the force rate is also negative, but only proportional to $-a$.

6.14 (a) $dF/da = 3Ka/2 - \sqrt{27\pi\varpi Ka^2/8}$; (b) $d\delta/da = 2a/R$; (c) $dF/d\delta = 3Ka/2 - \sqrt{27\pi\varpi KR^2/32a}$; (d) $k_i = -dF/d\delta$; (e) greater than zero.

6.15 Negative, because $W = \mathbf{f} \cdot \Delta\delta = |\mathbf{f}||\Delta\delta| \cos\varphi$, where φ is the angle between $\mathbf{f}$ and $\Delta\delta$. $\mathbf{f}$ points radially; $\Delta\delta$ points up, such that $\cos\varphi$ is negative.

6.16 (a) $dA = 2\pi R^2 \sin\theta\, d\theta$, (b) $dW = -2\pi R^2 |\mathbf{f}_A||\Delta\delta| \sin\theta \cos\theta\, d\theta$, (c) Triangular areas depend on both the differential area (greater at deeper indentation depths) and how much of the force is antiparallel to the change in depth (smaller at deeper depths). The largest triangular area will correspond to the maximum of $\sin\theta\cos\theta$, which is at 45°. (d) $W = -\pi R^2 |\mathbf{f}_A||\Delta\delta| \sin^2\theta_0$.

6.17 (a) GigaPascal, (b) 1.0 GPa, (c) 11 nm.

6.18 (a) MegaPascal, (b) 45 kPa, (c) 120 nm.

6.19 (a) $2\pi R^2(1 - \cos\theta_0) = 79$ nm^2, (b) 1,300 pairs, (c) 4.8 eV.

6.20 (a) ∞, (b) For a 70 kg person walking at 1.4 m/s, $\lambda = 6.76 \times 10^{-36}$ m, (c) No.

6.21 (a) 21.3 nm, (b) 0.497 nm, (c) 0.249 nm, (d) point particle, 0.11 nm, 0.28 nm (e) wave, wave, particle, respectively.

6.22 (a) 497 pm, (b) 144 pm, (c) 73.0 pm, (d) 47.8 pm. (e) 0.833 fm, 91.4 pm, 210 pm. (f) wave, wave, particle, particle, respectively.

6.23 (a) $F_{LJ}(r) = (12U_0/r_0)[(r_0/r)^{13} - (r_0/r)^7]$;
(b) $k_{LJ}(r) = (12U_0/r_0^2)[13(r_0/r)^{14} - 7(r_0/r)^8]$;
(c) r_0; (d) $-U_0$, $72U_0/r_0^2$, (e) $(-U_0, 72U_0/r_0^2)$ are the coordinates of the minimum.

6.24 (a) 1.64 pm, (b) 0.83 pm.

6.25 (a) 753 m/s, (b) 238 m/s, (c) 28.8 m/s.

6.26 (a) 1.88×10^{14} m/s, (b) 1.1 ps.

6.27 Typical research questions concern adhesion or elastic modulus.

6.28 Pre-Lab Question. Consult with your instructor.

6.29 Lab question. Comparing contact and dynamic modes, the relative heights of features. The phase difference in phase imaging (greater change for more compliant, stickier material). The rapidity of turnaround of a force curve (more quickly for stiffer material. The amount of adhesion as measured by a force curve (greater for more compliant, stickier material).

6.30 *Challenge Problem.*
(a) $\mathbf{E} = -\frac{1}{4\pi\varepsilon\varepsilon_0}\frac{q_1 a_1}{[(a_1/2)^2+r^2]^{3/2}}\hat{\mathbf{j}}$;

(b) $q_2 = \frac{1}{4\pi\varepsilon\varepsilon_0}\frac{\chi q_1 a_1/a_2}{[(a_1/2)^2+r^2]^{3/2}}$;

(c) $\mathbf{F}_{2on1} = \frac{q_1 q_2}{2\pi\varepsilon\varepsilon_0}[\frac{1}{r^2} - \frac{r}{(a^2+r^2)^{3/2}}]\hat{\mathbf{i}}$;

(d) $\mathbf{F}_{2on1} = \frac{1}{(4\pi\varepsilon\varepsilon_0)^2}\frac{2\chi q_1^2}{(r^2+(a/2r)^2)^{3/2}}[\frac{1}{r^2} - \frac{r}{(r^2+a^2)^{3/2}}]\hat{\mathbf{i}}$;

(e) $\mathbf{F}_{2on1} = \frac{3\chi(q_1 a)^2}{(4\pi\varepsilon\varepsilon_0)^2}\frac{1}{r^7}\hat{\mathbf{i}}$;

(f) $\mathbf{F}_{1on2} = -\frac{3\chi(q_1 a)^2}{(4\pi\varepsilon\varepsilon_0)^2}\frac{1}{r^7}\hat{\mathbf{i}}$;

(g) $U_6 = -\frac{1}{2}\frac{\chi(q_1 a)^2}{(4\pi\varepsilon\varepsilon_0)^2}\frac{1}{r^6}$.

6.31 *Challenge Problem.* The answers to the first two columns of data involve setting $\bar{F} = 0$ and $\bar{\Delta} = 0$, respectively. The coordinates of the minimum force for DMT can be seen from setting $\bar{A} = 0$. For BCP and JKRS, they stem from $(d\bar{F}/d\bar{A})(d\bar{A}/d\bar{\Delta}) = 0$. For making and breaking contact for DMT and BCP, these occur at $\bar{A} = 0$. Because of adhesion hysteresis in the JKRS model, these are two different sets of coordinates for making and breaking contact, one at $\bar{\Delta} = 0$ (Making) and one where $(d\bar{F}/d\bar{A})(d\bar{A}/d\bar{\Delta}) = \infty$ (Breaking).

Model	$\bar{\Delta}(\bar{F} = 0)$	$\bar{F}(\bar{\Delta} = 0)$	$(\bar{\Delta}_{min}, \bar{F}_{min})$	$(\bar{\Delta}_{MB}, \bar{F}_{MB})$
DMT	−1.587	−2	(0, −2)	(0, −2)
BCP	−1.165	−1.225	(+0.480, −1.375)	(+1, −1)
JKRS	−1.102	−1.333	(+0.437, −1.500)	$(0, -1.333)_{\mathrm{M}}$ $(+0.908, -0.833)_{\mathrm{B}}$

Chapter 7 Answers

7.1 Gravity incurs a constant offset to the cantilever's position, so it can be ignored.

7.2 Substitute $c_1 \exp i(\omega t + \theta)$, $h_1 \exp i\omega t$, and $F_1 \exp i\omega t$ into Eq. 7.2, take time derivatives, find $c_1 (\exp i\theta)\ \hat{r}$, and multiply numerator and denominator by the complex conjugate of the denominator. Collect the real and imaginary terms in the numerators.

7.3 Force control implies $h_1 = 0$ in Eq. 7.2. Take the real and imaginary parts and substitute them into Eqs. 7.5. Simplify.

7.4 Factor out k_c from the denominators of Eqs. 7.7. Substitute ω_k^2 for k_c/m and $\omega_k Q$ for k_c/b_c.

7.5 (a) $c_1 = F_1/k_c, \tan\theta = 0; k_c c_1/F_1 = 1, \tan\theta = 0$.
(b) $c_1 = F_1/\omega_k b_c, \tan\theta = -\infty; k_c c_1/F_1 = Q, \tan\theta = -\infty$.
(c) $c_1 = 0, \tan\theta = 0; k_c c_1/F_1 = 0, \tan\theta = 0$.

7.6 Factor out k_c^2 from the numerator and denominator of the equation for c_1. Factor out k_c^2 from the denominator of the equation for $\tan\theta$. For both equations, substitute ω_k^2 for k_c/m and $\omega_k Q$ for k_c/b_c.

7.7 (a) $c_1 = h_1, \tan\theta = 0; c_1/h_1 = 1, \tan\theta = 0$.
(b) $c_1 = h_1\sqrt{1 + k_c b_c^2/m}, \tan\theta = -\omega_k m/b_c; c_1/h_1 = Q\sqrt{1 + 1/Q^2}, \tan\theta = -Q$.
(c) $c_1 = 0, \tan\theta = -\infty; c_1/h_1 = 0, \tan\theta = -\infty$.

7.8 Substitute the expressions for $\omega_k, \omega_a, \omega_+, \omega_-$ into the tangent expression in Eq. 7.8. Use binomial expansions for square roots and denominators, assuming that Q is large. Drop small terms.

7.9 Substitute $q_1 \exp i(\omega t + \theta)$ and $V_1 \exp i\omega t$ into Eq. 7.12, take time derivatives, find $q_1 (\exp i\theta)\, \hat{r}$, and multiply numerator and denominator by the complex conjugate of the denominator. Collect the real and imaginary terms in the numerators.

7.10 Substitute the real and imaginary parts of Eq. 7.12 into Eqs. 7.5 and simplify. Check the result by substituting the equivalent variables discussed in Sect. 7.1.4.1 into Eqs. 7.7.

7.11 Substitute the equivalent variables discussed in Sect. 7.1.4.1 into Eqs. 7.7 to obtain q_1 and the tangent of the phase angle for an LRC circuit in series.

7.12 Use Eq. 7.17 and $\cos^2(\omega t) = (1 + \cos(2\omega t))$.

7.13 Use $\cos(\omega t + \theta) = \cos\omega t \cos\theta - \sin\omega t \sin\theta$ and $\sin^2\omega t = (1 - \cos 2t)/2$ to yield $\langle P_{Im}\rangle = -\omega q_1 V_1 \sin\theta/2$. It makes sense because the two expressions for power are the negative of each other. Hence the power absorbed equals the power dissipated.

7.14 (a) $c_1 = F_1/(k_c + \Delta k), \tan\theta = 0$. (b) $c_1 = 0, \tan\theta = 0$.
(c) $c_1 = F_1/(b_c + \Delta b)/\sqrt{(k_c + \Delta k)/m}, \tan\theta = -\infty$.

7.15 Lower than the frequency of B&E, single-valued; at the B&E frequency, double-valued; greater than the B&E frequency, triple-valued.

7.16 For the bistable curves in Fig. 7.10: Below resonance, the jumps from a round point to another, the larger one upon retraction; at resonance, equal sized jumps from a square point to another; above resonance, the jumps from a diamond point to another, the larger one upon retraction.

7.17 The amplitudes corresponding to hysteresis in the curves.

7.18 You will see that there are different protocols depending on the manufacturer.

7.19 **Lab question.** Confer with your lab instructor.

7.20 *Challenge Problem.* Follow the description in the text after Eq. 7.11. Use binomial expansions for square roots and denominators, assuming that Q is large. Drop small terms.

7.21 *Challenge Problem.* Follow the description in the text after Eq. 7.27.

Index

R

S

T

V

W

www.ingramcontent.com/pod-product-compliance
Lightning Source LLC
LaVergne TN
LVHW010336220926
848484LV00004BB/909